Solutions Manual to Accompany
Applied Survival Analysis
Regression Modeling of Time to Event Data

WILEY SERIES IN PROBABILITY AND STATISTICS

Established by WALTER A. SHEWHART and SAMUEL S. WILKS

Editors: *Peter Bloomfield, Noel A. C. Cressie, Nicholas I. Fisher, Iain M. Johnstone, J. B. Kadane, Louise M. Ryan, David W. Scott, Bernard W. Silverman, Adrian F. M. Smith, Jozef L. Teugels;*
Editors Emeriti: *Vic Barnett, Ralph A. Bradley, J. Stuart Hunter, David G. Kendall*

A complete list of the titles in this series appears at the end of this volume.

Solutions Manual to Accompany

Applied Survival Analysis
Regression Modeling of Time to Event Data

DAVID W. HOSMER, Jr.
STANLEY LEMESHOW

Solutions Manual Authored by
SUNNY KIM

WILEY-INTERSCIENCE
A JOHN WILEY & SONS, INC., PUBLICATION

This book is printed on acid-free paper.

Copyright © 2002 by John Wiley & Sons, Inc., New York. All rights reserved.

Published simultaneously in Canada.

No part of this publication may be reproduced, stored in a retrieval system or transmitted in any form or by any means, electronic, mechanical, photocopying, recording, scanning or otherwise, except as permitted under Sections 107 or 108 of the 1976 United States Copyright Act, without either the prior written permission of the Publisher, or authorization through payment of the appropriate per-copy fee to the Copyright Clearance Center, 222 Rosewood Drive, Danvers, MA 01923, (978) 750-8400, fax (978) 750-4744. Requests to the Publisher for permission should be addressed to the permissions Department, John Wiley & Sons, Inc., 605 Third Avenue, New York, NY 10158-0012, (212) 850-6011, fax (212) 850-6008, E-Mail: PERMREQ @ WILEY.COM.

For ordering and customer service, call 1-800-CALL-WILEY

Library of Congress Cataloging in Publication Data:

ISBN 0-471-24979-3 (paper)

Printed in the United States of America

10 9 8 7 6 5 4 3 2 1

CONTENTS

Preface		vii
Chapter 1	**Introduction to Regression Modeling of Survival Data**	1
	Exercise 1, 1	
	Exercise 2, 5	
	Exercise 3, 6	
Chapter 2	**Descriptive Methods for Survival Data**	9
	Exercise 1, 9	
	Exercise 2, 15	
	Exercise 3, 17	
	Exercise 4, 20	
	Exercise 5, 22	
	Exercise 6, 23	
	Exercise 7, 30	
	Exercise 8, 30	
Chapter 3	**Regression Models for Survival Data**	31
	Exercise 1, 31	
	Exercise 2, 36	
	Exercise 3, 36	
Chapter 4	**Interpretation of a Fitted Proportional Hazards Regression Model**	43
	Exercise 1, 43	
	Exercise 2, 48	
	Exercise 3, 52	
Chapter 5	**Model Development**	56
	Exercise 1, 56	
	Exercise 2, 61	
	Exercise 3, 71	

Chapter 6   Assessment of Model Adequacy   73

    Exercise 1,   73
    Exercise 2,   86
    Exercise 3,   88
    Exercise 4,   95
    Exercise 5,   106

Chapter 7   Extensions of the Proportional Hazards Model   119

    Exercise 1,   119
    Exercise 2,   123
    Exercise 3,   129
    Exercise 4,   132

Chapter 8   Parametric Regression Models   141

    Exercise 1,   141
    Exercise 2,   144
    Exercise 3,   148
    Exercise 4,   153
    Exercise 5,   155
    Exercise 6,   156
    Exercise 7,   181
    Exercise 8,   185

Chapter 9   Other Models and Topics   188

    Exercise 1,   188
    Exercise 2,   206
    Exercise 3,   213
    Exercise 4,   214
    Exercise 5,   219
    Exercise 6,   228

PREFACE

The problems at the end of each chapter of *Applied Survival Analysis: Regression Modeling of Time to Event Data* are designed to illustrate the methods, ideas, and approaches to analysis described in that chapter. While there are relatively few problems, each one may require a substantial amount of work to complete the required analysis. As such, the problems do not lend themselves to providing a partial solutions manual for students and a complete solutions manual for instructors. The format and approach of this manual is the same as that used for the solutions manual prepared by Elizabeth Donohoe-Cook for *Applied Logistic Regression, Second Edition* (Hosmer and Lemeshow, John Wiley & Sons, Inc., 2000).

We suggest that instructors supplement the problems in the text with similar questions using data from their own fields and areas of statistical practice. We are always interested in seeing new data that provide good teaching examples and sample data sets for additional exercises and exams. If you have such a data set please contact us and we will discuss including it in the archive of statistical data sets at the University of Massachusetts:

```
http://www-unix.oit.umass.edu/~statdata.
```

This solutions manual presents the methods, computer output, and discussion that we would use if we had been assigned the problems in the text. In any data analysis exercise one makes choices along the way, for example, which variables to include and how to scale continuous covariates. There is both art and science in a good data analysis and two experienced analysts may arrive at slightly different models, each of which accomplishes the goals of the analysis. Thus, in many problems our solution should not be taken as the only one possible. We encourage instructors to consider alternative solutions and models and to discuss their respective strengths and weaknesses with their students.

We performed almost all the calculations presented in this solutions manual using STATA (versions 6.0 and 7.0). The STATA code presented in the manual is what we used to complete the task and likely is not the most elegant or efficient coding. We have made no attempt to use all the "latest" features in the software. In addition, we have no plans to revise the solutions manual to illustrate future developments in STATA or any other package. Virtually all the calculations performed here in STATA are either available or can be programmed in other packages. We use SAS to obtain multivariate score and Wald tests.

We have made every attempt to make the solutions as accurate as possible. There is a formidable amount of numerical computation and calculation in the manual and there may be a few errors we missed. We would appreciate learning of these. However, unfortunately we do not have the time to react to and comment on alternative solutions to the problems.

As noted in the Preface of the text all data sets may be found at the Wiley ftp site:

```
ftp://ftp.wiley.com/public/sci_tech_med/logistic,
```

or on the data set archive whose URL is shown above.

DAVID W. HOSMER, JR.
STANLEY LEMESHOW
SUNNY KIM

Chapter One – Solutions

1. *Using the data from the Worcester Heart Attack Study:*

    (a) Graph length of follow-up versus age using the censoring variable at follow-up as the plotting symbol for each of the pooled cohorts defined by YRGRP. Are the plots basically the same or do they differ in shape in an important way? Is it possible to tell from the shape of the plot if age is a predictor of survival time?

    ```
    . use whas.dta
    . graph lenfoll age if yrgrp==1, s([censor]) ylab xlab saving(p1,replace) trim(1) psize(125)
    . graph lenfoll age if yrgrp==2, s([censor]) ylab xlab saving(p2,replace) trim(1) psize(125)
    . graph lenfoll age if yrgrp==3, s([censor]) ylab xlab saving(p3,replace) trim(1) psize(125)
    . graph using p1 p2 p3
    ```

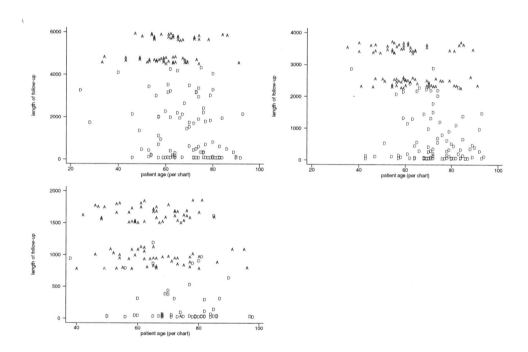

The shapes of all three plots are basically the same. Since the cloud is weakly trapezoidal in a negative direction, age could be a predictor of survival time. While hard to see in the plot, the censored observations are plotted with an "A" and tend to fall above the non-censored observations plotted with a "D". The bands in the plots reflect that for censored observations the follow-up time is determined at somewhat regular intervals, e.g. 3 or 6 months.

    (b) What key characteristics of the data plotted in problem 1(a) should be kept in mind when choosing a possible regression model?

Censoring, positive values, right skewness.

(c) *By eye, draw on each of the three scatterplots from problem 1(a) what you feel is the best regression function for a survival time regression model.*

It would begin in the top left corner and drop sharply, curving to the lower right

(d) *Obtain a cross tabulation of YRGRP and the censoring variable FSTAT and compute the percent dead and the percent censored in each of the three groups. What effect do you think the difference in the percent censored should have on the location of the lines drawn in problem 1(c)?*

```
. tab yrgrp fstat, row

          | follow-up status as
          |    of last date
    yrgrp |    Alive      Dead |     Total
----------+----------------------+----------
    75-78 |       58       102 |       160
          |    36.25     63.75 |    100.00
----------+----------------------+----------
    81-84 |       82        93 |       175
          |    46.86     53.14 |    100.00
----------+----------------------+----------
    86-88 |       92        54 |       146
          |    63.01     36.99 |    100.00
----------+----------------------+----------
    Total |      232       249 |       481
          |    48.23     51.77 |    100.00
```

Since the value of the censored observations represent a lower bound on unobserved survival times, a higher percentage of censored observations should shift the curve upward. The percentage of censored observations in year 81-84 and 86-88 are 47% and 63%, respectively. Since the percentage of censored observation is larger for the year 86-88, its curve will be shifted to upward (equivalently, larger β_0) compared to the year 81-84.

(e) *Fit the exponential regression model to the data in each of the three scatterplots and add the fitted values to the plot (e.g., see Figure 1(d)). How do the regression fitted values compare to the ones drawn in problem 1(c)? Is the response to problem 1(d) correct?*

(i) YRGRP=1

```
. streg age if yrgrp==1,dist(exp) nohr time nolog

        failure _d:  fstat
   analysis time _t: lenfoll


Exponential regression -- accelerated failure-time form

No. of subjects =          160              Number of obs   =        160
No. of failures =          102
Time at risk    =       415690
                                            LR chi2(1)      =      30.99
Log likelihood  =   -404.10817              Prob > chi2     =     0.0000

------------------------------------------------------------------------------
         _t |      Coef.   Std. Err.       z    P>|z|     [95% Conf. Interval]
------------+----------------------------------------------------------------
        age |  -.0483987   .0087456    -5.53   0.000    -.0655398   -.0312577
      _cons |   11.49938   .6113469    18.81   0.000     10.30116    12.69759
------------------------------------------------------------------------------

. predict that1 if yrgrp==1, mean time
```

(ii) YRGRP=2

```
. streg age if yrgrp==2,dist(exp) nohr time nolog

        failure _d:  fstat
   analysis time _t: lenfoll


Exponential regression -- accelerated failure-time form

No. of subjects =          175              Number of obs   =        175
No. of failures =           93
Time at risk    =       292301
                                            LR chi2(1)      =      50.50
Log likelihood  =   -365.81805              Prob > chi2     =     0.0000

------------------------------------------------------------------------------
         _t |      Coef.   Std. Err.       z    P>|z|     [95% Conf. Interval]
------------+----------------------------------------------------------------
        age |  -.0573122   .0082075    -6.98   0.000    -.0733987   -.0412257
      _cons |    11.9492   .6057967    19.72   0.000     10.76186    13.13654
------------------------------------------------------------------------------

. predict that2 if yrgrp==2, mean time
```

(iii) YRGRP=3

```
. streg age if yrgrp==3,dist(exp) nohr time nolog

         failure _d:  fstat
   analysis time _t:  lenfoll


Exponential regression -- accelerated failure-time form

No. of subjects =          146                Number of obs   =        146
No. of failures =           54
Time at risk    =       126564
                                              LR chi2(1)      =      27.31
Log likelihood  =   -272.93264                Prob > chi2     =     0.0000

------------------------------------------------------------------------------
         _t |      Coef.   Std. Err.      z    P>|z|     [95% Conf. Interval]
------------+-----------------------------------------------------------------
        age |  -.0616236   .0118895    -5.18   0.000    -.0849265   -.0383207
      _cons |   12.06678   .8902814    13.55   0.000     10.32186     13.8117
------------------------------------------------------------------------------

. predict that3 if yrgrp==3, mean time
```

```
. sort age
. graph that1 lenfoll age if yrgrp==1, s(i[fstat]) ylab xlab saving(p1,replace)
trim(1) psize(125)c(1)
. graph that2 lenfoll age if yrgrp==2, s(i[fstat]) ylab xlab saving(p2,replace)
trim(1) psize(125) c(1)
. graph that3 lenfoll age if yrgrp==3, s(i[fstat]) ylab xlab saving(p3,replace)
trim(1) psize(125) c(1)
```

`. graph using p1 p2 p3`

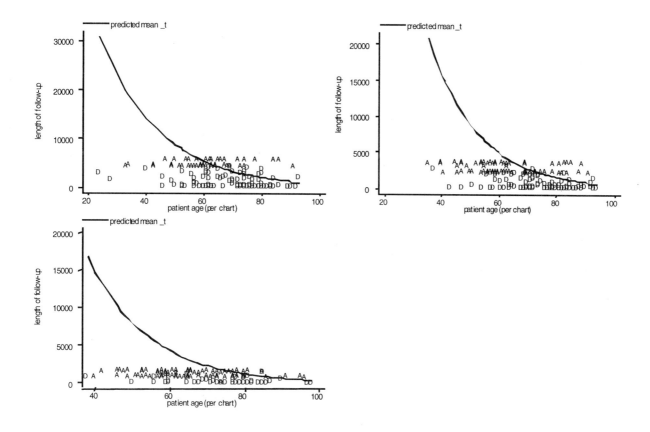

Note that the estimate of β_0 increases slightly over the three grouped cohorts and due to the high percent of censoring the fitted lines lie well above the scatterplot.

2. *What key characteristics about the observations of total length of follow-up must be kept in mind when considering the computation of simple univariate descriptive statistics?*

Right censoring of the data and skewness

6 CHAPTER 1 SOLUTIONS

3. *Repeat problems 1 and 2 using time to return to drug use and age in the UIS and grouping by study site.*

   (a) *Graph the time to return to drug use and age by study site. Are the plots basically the same or do they differ in shape in an important way? Is it possible to tell from the shape of the plot if age is a predictor of survival time?*

```
. use uis
. sort age
. graph time age if site==0, s([censor]) ylab xlab saving(p1,replace) trim(1)
  psize(125)
. graph time age if site==1, s([censor]) ylab xlab saving(p2,replace) trim(1)
  psize(125)
. graph using p1 p2
```

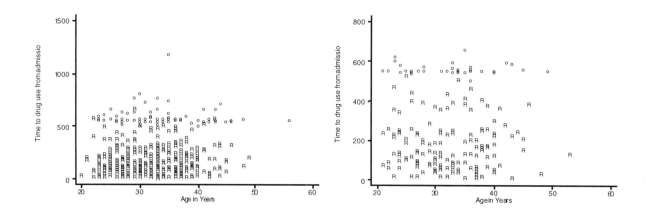

The shapes of two plots are basically the same. Since the cloud is rectangular without obvious trends, it is hard to tell that if age is a predictor of the time to return to drug use.

   (c) *By eye, draw on each of the three scatterplots from problem 3(a) what you feel is the best regression function for a survival time regression model.*

It would be close to a horizontal line.

(d) *Obtain a cross tabulation of SITE and the censoring variable CENSOR and compute the percent dead and the percent censored in each site.*

```
. tab site censor, row

          |     censor
     site |        0          1 |    Total
----------+----------------------+----------
        0 |       80        364 |      444
          |    18.02      81.98 |   100.00
----------+----------------------+----------
        1 |       40        144 |      184
          |    21.74      78.26 |   100.00
----------+----------------------+----------
    Total |      120        508 |      628
          |    19.11      80.89 |   100.00
```

Since the percentage of censored observation in site = 1 is just slightly larger than site = 0, the curve for site = 1 will be shifted slightly upward (equivalently, larger β_0) compared to the site = 0.

(e) *Fit the exponential regression model to the data in each of the three scatterplots and add the fitted values to the plot (e.g., see Figure 3(d)). How do the regression fitted values compare to the ones drawn in problem 3(c)? Is the response to problem 3(d) correct? .*

(i) Site=0

```
. streg age if site==0,dist(exp) nohr time nolog

         failure _d:  censor
   analysis time _t:  time

Exponential regression -- accelerated failure-time form

No. of subjects =         439                Number of obs    =        439
No. of failures =         360
Time at risk    =      100703
                                             LR chi2(1)       =       7.08
Log likelihood  =   -729.89446               Prob > chi2      =     0.0078

------------------------------------------------------------------------------
         _t |      Coef.   Std. Err.      z    P>|z|     [95% Conf. Interval]
------------+----------------------------------------------------------------
        age |   .0231842   .0088178     2.63   0.009     .0059018    .0404667
      _cons |    4.87695   .2889746    16.88   0.000     4.31057     5.44333
------------------------------------------------------------------------------

. predict that0 if site==0, mean time
(189 missing values generated)
```

(ii) Site=1

```
. streg age if site==1,dist(exp) nohr time nolog

        failure _d:  censor
   analysis time _t: time


Exponential regression -- accelerated failure-time form

No. of subjects =          184                Number of obs   =        184
No. of failures =          144
Time at risk    =        46113
                                              LR chi2(1)      =       0.05
Log likelihood  =   -292.75344                Prob > chi2     =     0.8307

------------------------------------------------------------------------------
         _t |     Coef.   Std. Err.      z    P>|z|     [95% Conf. Interval]
-------------+----------------------------------------------------------------
        age |  -.002707   .0126487    -0.21   0.831    -.027498    .022084
      _cons |  5.855709   .4141939    14.14   0.000     5.043904    6.667514
------------------------------------------------------------------------------

. predict that1 if site==1, mean time
(444 missing values generated)
```

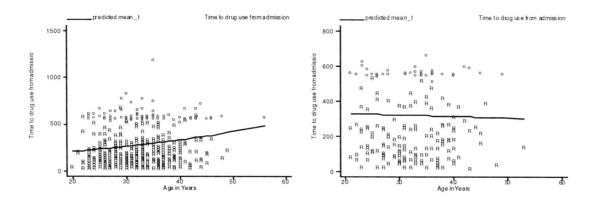

Note that the estimate of the intercept for site = 1 is larger and age is significant only in site = 0.

Chapter Two – Solutions

1. Listed below are values of survival time (length of follow-up) for 6 males and 6 females from the WHAS. Right-censored times are denoted by a "+" as a superscript
   Males: 1, 3, 4+, 10, 12, 18
   Females: 1, 3+, 6, 10, 11, 12+
   Using these data, compute by hand (and verify hand calculations when possible using STATA) the following:
   (a) The Kaplan-Meier estimate of the survivorship function for each gender.
   (b) Compute the pointwise 95 percent confidence intervals for the survivorship functions estimated in problem 1(a).

Males:

Time	di	ni	Shat(t)
0	0	6	1.0000
1	1	6	0.8333
3	1	5	0.6667
4	0	4	0.6667
10	1	3	0.4444
12	1	2	0.2222
18	1	1	0.0000

Females:

Time	di	ni	Shat(t)
0	0	6	1.0000
1	1	6	0.8333
3	0	5	0.8333
6	1	4	0.6250
10	1	3	0.4167
11	1	2	0.2083
12	0	1	0.2083

```
. input sex time censor
            sex         time       censor
  1.  1 1 1
  2.  1 3 1
  3.  1 4 0
  4.  1 10 1
  5.  1 12 1
  6.  1 18 1
  7.  2 1 1
  8.  2 3 0
  9.  2 6 1
 10.  2 10 1
 11.  2 11 1
 12.  2 12 0
 13. end
```

```
. label define sexlbl 1 male 2 female
. label values sex sexlbl
. sts list, by (sex)

         failure _d:  censor
   analysis time _t:  time

              Beg.              Net      Survivor    Std.
   Time      Total    Fail      Lost     Function   Error     [95% Conf. Int.]
   ---------------------------------------------------------------------------
   male
      1         6       1         0       0.8333    0.1521    0.2731   0.9747
      3         5       1         0       0.6667    0.1925    0.1946   0.9044
      4         4       0         1       0.6667    0.1925    0.1946   0.9044
     10         3       1         0       0.4444    0.2222    0.0662   0.7849
     12         2       1         0       0.2222    0.1925    0.0096   0.6147
     18         1       1         0       0.0000      .         .        .
   female
      1         6       1         0       0.8333    0.1521    0.2731   0.9747
      3         5       0         1       0.8333    0.1521    0.2731   0.9747
      6         4       1         0       0.6250    0.2135    0.1419   0.8931
     10         3       1         0       0.4167    0.2218    0.0560   0.7665
     11         2       1         0       0.2083    0.1844    0.0087   0.5951
     12         1       0         1       0.2083    0.1844    0.0087   0.5951
   ---------------------------------------------------------------------------
```

1(a): Equation (2.1) on page 34 is used for hand calculation

1(b): (i) Greenwood estimate of standard error, the equation (2.5) on page 42, is used and is shown in the standard error column above.

(ii) Confidence interval estimation based on log-log survivorship function, equations (2.6)-(2.8) on page 43 – 44, are used and shown in the last two columns above.

(c) The Hall and Wellner 95% confidence bands for the survivorship functions estimated in problem 1(a).

Stata does not calculate Hall and Wellner confidence bands automatically. Equations (2.9) and (2.10) on page 45 may be used for calculating these confidence bands.

```
.sts gen riskset=n, by( sex)
.sts gen no_died=d, by( sex)
.sts generate survcurv=s, by( sex)
.sts gen se_s=se(s), by( sex)
.sts generate pointlb=lb, by( sex)
.sts generate pointub=ub, by( sex)
.list riskset no_died survcurv se_s pointlb pointub if sex==2
.list riskset no_died survcurv se_s pointlb pointub if sex==1
.gen var_tm=( se_s^2)/( survcurv^2)
```

```
.list riskset no_died survcurv se_s   var_tm if sex==1
.list riskset no_died survcurv se_s   var_tm if sex==2
.gen ahat=(6* var_tm)/(1+6* var_tm)
.list riskset no_died survcurv se_s   var_tm  ahat if sex==1
.list riskset no_died survcurv se_s   var_tm  ahat if sex==2
.gen lb=ln(-ln( survcurv))-1.356*((1+6* var_tm)/(sqrt(6)*abs(ln( survcurv))))
.gen ub=ln(-ln( survcurv))+1.356*((1+6* var_tm)/(sqrt(6)*abs(ln( survcurv))))
.gen hw_lb=exp(-exp(ub))
.gen hw_ub=exp(-exp(lb))

.list riskset survcurv se_s   var_tm   hw_lb hw_ub if sex==1
.list riskset survcurv se_s   var_tm   hw_lb hw_ub if sex==2
.list pointlb pointub hw_lb hw_ub survcurv time if sex==1
.list pointlb pointub hw_lb hw_ub survcurv time if sex==2

.graph pointlb pointub hw_lb hw_ub survcurv time if sex==1, s(ppooO) c(JJJJJ) ylab
  xlab
.graph pointlb pointub hw_lb hw_ub survcurv time if sex==2, s(ppooO) c(JJJJJ) ylab
  xlab
```

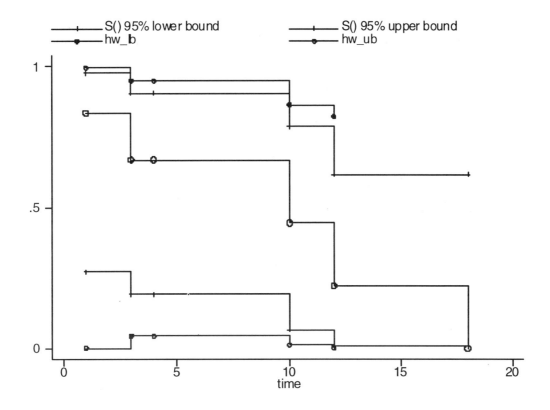

(d)   Point and 95 percent confidence interval estimates of the 25th, 50th and 75th percentiles of survival time distribution for each gender.

Note: In Stata, the *p*-percentile of survival time is the analysis time at which *p*% of subjects have failed and (100-*p*)% have not.

```
. stsum , by(sex)

        failure _d:  censor
     analysis time _t:  time

                             incidence       no. of    |------ Survival time -----|
sex         | time at risk      rate        subjects        25%        50%        75%
------------+---------------------------------------------------------------------
      male  |       48        .1041667          6            3         10         12
    female  |       43        .0930233          6            6         10         11
------------+---------------------------------------------------------------------
     total  |       91        .0989011         12            6         10         12
```

Please note:

STATA's stci command reports an upper limit for the confidence interval that is one ranked survival time larger than the methods described in the text. The rationale is to obtain a slightly wider confidence interval that may have better coverage properties. In this solutions manual we report the smaller value consistent with the text. To our knowledge no simulations have been performed that compares the two choices for an upper limit. Other packages (e.g., SAS) base the choice of endpoints on the estimated survival function and its standard error. This can lead to different results due to the non-linear relationship between endpoints based on the log-log and survivorship functions.

```
. stci, by(sex) p(25)  noshow              estimates the time 75% are alive

              no. of
sex         | subjects     25%      Std. Err.    [95% Conf. Interval]
------------+---------------------------------------------------------
         0  |      6        3         1.5            1         12
         1  |      6        6        2.10256         1         11
------------+---------------------------------------------------------
     total  |     12        3        3.28125         1         10
```

The correct upper endpoint is 10 for both groups.

```
. stci, by(sex) p(50) noshow

              no. of
sex         | subjects     50%      Std. Err.    [95% Conf. Interval]
------------+---------------------------------------------------------
      male  |      6       10       4.250364         1          .
    female  |      6       10       3.022481         1          .
------------+---------------------------------------------------------
     total  |     12       10       2.938524         1         12
```

```
. stci, by(sex) p(75) noshow

            |    no. of
    sex     |   subjects           75%      Std. Err.      [95% Conf. Interval]
------------+-----------------------------------------------------------------
    male    |         6             12       2.393048          3          .
    female  |         6             11          .              6          .
------------+-----------------------------------------------------------------
    total   |        12             12       2.895292         10          .
```

(e) *The mean survival time for each gender using all available times.*

Male

Time	Di	ni	Shat(t)	t(i)-t(i-1)	areas
0	0	6	1	1	1
1	1	6	0.833	2	1.6667
3	1	5	0.667	7	4.6667
4	0	4	0.667		
10	1	3	0.444	2	0.8889
12	1	2	0.222	6	1.3333
18	1	1	0		
				mean	**9.5556**

Female

Time	di	ni	Shat(t)	t(i)-t(i-1)	Areas
0	0	6	1	1	1
1	1	6	0.833	5	4.16667
3	0	5	0.833		
6	1	4	0.625	4	2.5
10	1	3	0.417	1	0.41667
11	1	2	0.208	1	0.20833
12	0	1	0.208		
				mean	**8.29167**

```
. stci, rmean by(sex)      rmean is calculated as the area under the KM curve

        failure _d:  censor
  analysis time _t:  time

            |    no. of     restricted
    sex     |   subjects        mean       Std. Err.     [95% Conf. Interval]
------------+-----------------------------------------------------------------
    male    |         6      9.555556      2.554348       4.54913     14.562
    female  |         6      8.291667(*)   1.634897       5.08733     11.496
------------+-----------------------------------------------------------------
    total   |        12      9.52381       1.753523       6.08697     12.9607

(*)largest observed analysis time is censored, mean is underestimated
```

14 CHAPTER 2 SOLUTIONS

When the largest observation time is censored, stci "emean" option will extend the survivor function from the last observed time to zero using an exponential function and will compute the area under the entire curve.

```
. stci, emean by(sex)

        failure _d:  censor
   analysis time _t: time

              |   no. of     extended
         sex  |  subjects       mean
   -----------+-------------------------
         male |      6       9.555556(*)
       female |      6       9.885428
   -----------+-------------------------
        total |     12       9.52381(*)

(*) no extension needed
```

(f) *A graph of the estimated survivorship functions for each gender computed in problem 1(a) along with the pointwise 95 percent limit computed in problems 1(b).*

`. sts graph, gwood by(sex)`

Kaplan-Meier survival estimates, by sex
95%, pointwise confidence band shown

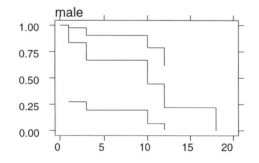
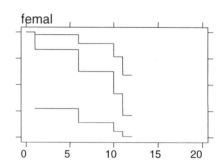

time

Graphs by sex

2. *Repeat problem 1 using data from grouped cohort 1 (1975 and 1978) from the Worcester Heart Attack Study.*
   (a)   The Kaplan-Meier estimate of the survivorship function for each gender.
   (b)   Pointwise 95 percent confidence intervals for the survivorship functions estimated in problem 2(a).
   (f)   A graph of the estimated survivorship functions for each gender computed in problem 2(a) along with the pointwise and overall 95% limit computed in problems 2(b) and 2(c).

```
. sts list if yrgrp==1, by(sex)    Displays numeric value of KM estimates output
extensive and not shown here.

. sts graph if  yrgrp==1,by(sex) gwood
```

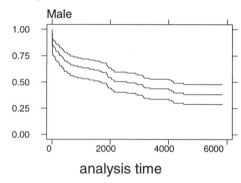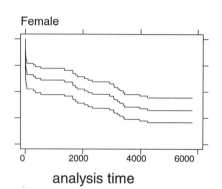

Graphs by patient gender

   (c)   The Hall and Wellner 95 percent confidence bands for the survivorship functions estimated in problem 2(a).

Output not shown; but the STATA code shown in problem 1(c) can be adapted to the current problem

(d) *Point and 95 percent confidence interval estimates of the 25th, 50th and 75th percentiles of survival time distribution for each gender.*

```
. stsum  if yrgrp==1, by(sex)

       failure _d:  fstat
   analysis time _t:  lenfol

          |                 incidence      no. of   |------ Survival time -----|
   sex    | time at risk      rate       subjects  |    25%       50%       75%
----------+-------------------------------------------------------------------
    Male  |     270785      .000229          101        326      2564        .
  Female  |     144905      .000276           59         13      2187        .
----------+-------------------------------------------------------------------
   total  |     415690     .0002454          160        144      2335        .
```

```
. stci if yrgr==1, by( sex) p(25) noshow

          |    no. of
   sex    |  subjects       25%       Std. Err.     [95% Conf. Interval]
----------+-------------------------------------------------------------
    Male  |      101        326       76.69914          60         623
  Female  |       59         13        5.318048          3         346
----------+-------------------------------------------------------------
   total  |      160        144       64.32257          14         410
```

Correct upper endpoints are 618 and 281

```
. stci if yrgr==1, by( sex) p(50) noshow

          |    no. of
   sex    |  subjects       50%       Std. Err.     [95% Conf. Interval]
----------+-------------------------------------------------------------
    Male  |      101       2564       285.7682        1696        4162
  Female  |       59       2187       290.7548         346        3402
----------+-------------------------------------------------------------
   total  |      160       2335       285.3502        1748        3280
```

Correct upper endpoints are 4105 and 3280

```
. stci if yrgr==1, by( sex) p(75) noshow

          |    no. of
   sex    |  subjects       75%       Std. Err.     [95% Conf. Interval]
----------+-------------------------------------------------------------
    Male  |      101          .              .             .           .
  Female  |       59          .              .          3441           .
----------+-------------------------------------------------------------
   total  |      160          .              .             .           .
```

(e) *The mean survival time for each gender using all available times.*

```
. stci if yrgr==1, rmean by( sex)   noshow          restricted mean

             |  no. of    restricted
    sex      | subjects         mean    Std. Err.    [95% Conf. Interval]
-------------+----------------------------------------------------------
    Male     |    101     2989.572(*)   247.9831     2503.53    3475.61
    Female   |     59     2664.068(*)   319.9238     2037.03    3291.11
-------------+----------------------------------------------------------
    total    |    160     2873.99(*)    196.8172     2488.24    3259.74

(*) largest observed analysis time is censored, mean is underestimated.
```

```
. stci if yrgr==1, emean by( sex)   noshow          extended mean

             |  no. of      extended
    sex      | subjects         mean
-------------+----------------------
    Male     |    101       5329.623
    Female   |     59       4314.455
-------------+----------------------
    total    |    160       4945.002
```

3. *Repeat problem 1 using grouped cohort 1(1975 and 1978) from the WHAS with four groups defined by the age intervals: [24,60], [61,65], [66,75], and [76,99]. In this subgroup of the data, 60, 65, and 75 are approximately the three quartiles of the age distribution.*

```
. gen agegroup = age if yrgrp==1
(321 missing values generated)

. recode agegroup  24/60=1 61/65=2 66/75=3 76/93=4
(160 changes made)
```

(a) *The Kaplan-Meier estimate of the survivorship function for each age interval.*

```
. sts list if yrgrp==1, by( agegroup)  Extensive output not shown. It displays numeric
value of KM estimates
```

(b) *Pointwise 95 percent confidence intervals for the survivorship functions.*

```
. sts list if yrgrp==1, by(agegroup)  Displays the pointwise confidence intervals
```

(c) *Hall and Wellner confidence bands.*

As for problem 2 the STATA code shown in problem 1(c) can be adapted to proved the Hall and Wellner bands.

(d) *Point and 95 percent confidence interval estimates of the 25th, 50th and 75th percentiles of survival time distribution for each age group.*

```
. stsum if yrgrp==1, by( agegroup) noshow

              |                 incidence    no. of    |------ Survival time -----|
     agegroup | time at risk       rate     subjects        25%        50%       75%
   -----------+-----------------------------------------------------------------------
            1 |      161238      .0001426        50        1031          .         .
            2 |      105692      .0001608        32         218       4162         .
            3 |      101897      .0002552        38         518       2564         .
            4 |       46863      .0007682        40           5        239      2060
   -----------+-----------------------------------------------------------------------
        total |      415690      .0002454       160         144       2335         .
```

```
. stci if yrgrp==1, by(agegroup) p(25) noshow

              |    no. of
     agegroup |   subjects        25%       Std. Err.     [95% Conf. Interval]
   -----------+------------------------------------------------------------------
            1 |        50        1031       282.8247         107        2187
            2 |        32         218       288.1519          16        3139
            3 |        38         518       180.2588          10        1653
            4 |        40           5       .1971699           3          14
   -----------+------------------------------------------------------------------
        total |       160         144       64.32257          14         410
```

Correct upper endpoints are 2050, 3078, 1614 and 13

```
. stci if yrgrp==1, by(agegroup) p(50) noshow

              |    no. of
     agegroup |   subjects        50%       Std. Err.     [95% Conf. Interval]
   -----------+------------------------------------------------------------------
            1 |        50           .              .        2050           .
            2 |        32        4162              .        1902           .
            3 |        38        2564       282.7545        1442        4240
            4 |        40         239       42.11952           6         623
   -----------+------------------------------------------------------------------
        total |       160        2335       285.3502        1748        3280
```

Correct upper endpoints for groups 3 and 4 are 3402 and 538.

```
. stci if yrgrp==1, by(agegroup) p(75) noshow

                |   no. of
       agegroup |  subjects        75%      Std. Err.       [95% Conf. Interval]
       ---------+----------------------------------------------------------------
              1 |        50          .            .              .           .
              2 |        32          .            .              .           .
              3 |        38          .            .           3280           .
              4 |        40       2060     122.4026            410        3207
       ---------+----------------------------------------------------------------
          total |       160          .            .              .           .
```

Correct upper endpoint for group 4 is 2936.

(e) The mean survival time for each agegroup

```
. stci if yrgrp==1, rmean by(agegroup) noshow

                |   no. of    restricted
       agegroup |  subjects        mean     Std. Err.       [95% Conf. Interval]
       ---------+----------------------------------------------------------------
              1 |        50     3715.9(*)    346.7384        3036.31     4395.49
              2 |        32   3576.938(*)    432.9544        2728.36     4425.51
              3 |        38   2881.224(*)    364.832         2166.17     3596.28
              4 |        40    1210.85(*)    291.005          640.491    1781.21
       ---------+----------------------------------------------------------------
          total |       160    2873.99(*)    196.8172        2488.24     3259.74

(*) largest observed analysis time is censored, mean is underestimated.
```

```
. stci if yrgrp==1, emean by(agegroup) noshow

                |   no. of     extended
       agegroup |  subjects        mean
       ---------+----------------------
              1 |        50     8836.463
              2 |        32     7185.583
              3 |        38     4381.234
              4 |        40     1462.437
       ---------+----------------------
          total |       160     4945.002
```

CHAPTER 2 SOLUTIONS

(f) *A graph of the estimated survivorship functions for each gender computed in problem 3(a) along with the pointwise and overall 95%limit computed in problems 3(b) and 3(c).*

```
. sts graph, gwood by(agegroup) bsize(120)
```

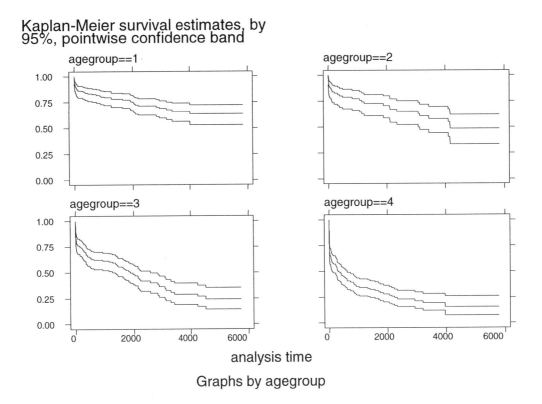

4. *Compute by hand, and verify hand calculations with a software package, the log-rank, generalized Wilcoxon, and Peto-Prentice tests for the equality of two survivorship functions estimated in problem 1(a).*

Here we only show the computer output.

```
. sts test sex          log-rank test is default

        failure _d:  censor
   analysis time _t:  time

Log-rank test for equality of survivor functions
------------------------------------------------
        |  Events
sex     |  observed      expected
--------+-------------------------
male    |     5            5.10
female  |     4            3.90
--------+-------------------------
Total   |     9            9.00

              chi2(1) =     0.00
              Pr>chi2 =     0.9437
```

```
. sts test sex,wilcoxon

     failure time:  time
     failure/censor:  censor
Wilcoxon (Breslow) test for equality of survivor functions
----------------------------------------------------------
        |  Events                      Sum of
sex     |  observed      expected      ranks
--------+----------------------------------------
1       |     5            5.10           1
2       |     4            3.90          -1
--------+----------------------------------------
Total   |     9            9.00           0
              chi2(1) =     0.01
              Pr>chi2 =     0.9281
```

```
. sts test sex, peto

        failure _d:  censor
   analysis time _t:  time

Peto-Peto test for equality of survivor functions

        |  Events        Events        Sum of
sex     |  observed      expected      ranks
--------+----------------------------------------
1       |     5            5.10        1.388e-17
2       |     4            3.90           0
--------+----------------------------------------
Total   |     9            9.00           0
              chi2(1) =     0.00
              Pr>chi2 =     1.0000
```

5. *Repeat problem 4 using data from grouped cohort 1 (1975 and 1978) of the WHAS. Do the results of the test support what is seen in the graphs of the estimated survivorship functions?*

Since the expected and observed numbers of observation are not significantly different, we fail to reject the null hypothesis. We conclude that the survival experience for males and females are not significantly different. This test supports what is seen in the graphs.

```
. sts test sex if yrgrp==1

        failure _d:  fstat
   analysis time _t:  lenfoll
Log-rank test for equality of survivor functions
------------------------------------------------
        |   Events
sex     | observed        expected
--------+-------------------------
Male    |     62            66.97
Female  |     40            35.03
--------+-------------------------
Total   |    102           102.00
             chi2(1) =      1.08
             Pr>chi2 =    0.2984
```

```
. sts test sex if yrgrp==1, wilcoxon

        failure _d:  fstat
   analysis time _t:  lenfol
Wilcoxon (Breslow) test for equality of survivor functions
----------------------------------------------------------
        |   Events                          Sum of
sex     | observed        expected           ranks
--------+-----------------------------------------
Male    |     62            66.97            -738
Female  |     40            35.03             738
--------+-----------------------------------------
Total   |    102           102.00               0
             chi2(1) =      1.85
             Pr>chi2 =    0.1741
```

```
. sts test   sex   if yrgrp==1, peto
         failure _d:  fstat
   analysis time _t: lenfol

Peto-Peto test for equality of survivor functions
          |   Events       Events      Sum of
  sex     | observed      expected      ranks
----------+--------------------------------------
  Male    |     62          66.97    -4.4613539
  Female  |     40          35.03     4.4613539
----------+--------------------------------------
  Total   |    102         102.00            0

             chi2(1) =       1.77
             Pr>chi2 =     0.1828
```

6. *Repeat problem 4 using data from grouped cohort 1 (1975 and 1978) of the WHAS with four groups defined by the age intervals: [24,60], [61,65], [66,75], and [76,93]. Using the midpoints of the four age intervals, test for trend using the test statistic defined in (2.28). In addition, test whether the survivorship experience for the middle two age groups is the same or different from the youngest and oldest age groups.*

   (i)   Test for the equality of four survivorship functions

Since each statistic is significant, we reject the hypothesis that the survivorship functions for the four age groups are the same.

```
. sts graph if  yrgrp==1, by( agegroup)
```

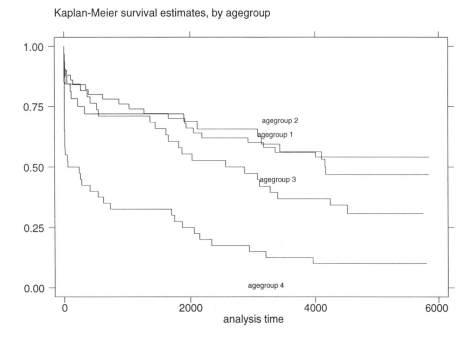

```
. sts test agegroup if yrgrp==1

        failure _d:  fstat
   analysis time _t:  lenfol
Log-rank test for equality of survivor functions
------------------------------------------------
          |  Events
 agegroup |  observed       expected
----------+-----------------------------
    1     |     23            37.51
    2     |     17            23.78
    3     |     26            25.25
    4     |     36            15.45
----------+-----------------------------
   Total  |    102           102.00
              chi2(3) =      36.02
              Pr>chi2 =       0.0000
```

```
. sts test agegroup if yrgrp==1, wilcoxon
        failure _d:  fstat
   analysis time _t:  lenfol

Wilcoxon (Breslow) test for equality of survivor functions
----------------------------------------------------------
          |  Events                         Sum of
 agegroup |  observed       expected         ranks
----------+-----------------------------------------
    1     |     23            37.51          -1549
    2     |     17            23.78           -807
    3     |     26            25.25           -166
    4     |     36            15.45           2522
----------+-----------------------------------------
   Total  |    102           102.00              0
              chi2(3) =      35.36
              Pr>chi2 =       0.0000
```

```
. sts test agegroup if yrgrp==1, peto
 failure _d:  fstat
   analysis time _t:  lenfol
Peto-Peto test for equality of survivor functions

          |   Events      Events       Sum of
 agegroup |  observed    expected        ranks
----------+---------------------------------------
    1     |      23        37.51     -9.5843937
    2     |      17        23.78     -4.9631422
    3     |      26        25.25     -.95336946
    4     |      36        15.45      15.500905
----------+---------------------------------------
 Total    |     102       102.00             0

              chi2(3) =     35.31
              Pr>chi2 =    0.0000
```

(ii)   Trend test

The midpoints of the four groups are (42, 63, 70.5, 84.5). Any linear transformation of this vector of coefficients would yield the same value of the test statistic.

Test result indicates strong evidence for a trend in survival experience.

```
. gen midpoint = agegroup
. recode midpoint  1=42 2=63 3=70.5 4=84.5
(481 changes made)
```

```
. sts test midpoint  if yrgrp==1, trend noshow

Log-rank test for equality of survivor functions

          |   Events      Events
 midpoint |  observed    expected
----------+------------------------
   42     |      23        37.51
   63     |      17        23.78
   70.5   |      26        25.25
   84.5   |      36        15.45
----------+------------------------
 Total    |     102       102.00

              chi2(3) =     36.02
              Pr>chi2 =    0.0000

Test for trend of survivor functions

              chi2(1) =     23.67
              Pr>chi2 =    0.0000
```

```
. sts test midpoint   if yrgrp==1, trend wilcoxon noshow
```

Wilcoxon (Breslow) test for equality of survivor functions

midpoint	Events observed	Events expected	Sum of ranks
42	23	37.51	-1549
63	17	23.78	-807
70.5	26	25.25	-166
84.5	36	15.45	2522
Total	102	102.00	0

```
            chi2(3) =     35.36
            Pr>chi2 =    0.0000
```

Test for trend of survivor functions

```
            chi2(1) =     22.74
            Pr>chi2 =    0.0000
```

```
. sts test midpoint   if yrgrp==1, trend peto noshow
```

Peto-Peto test for equality of survivor functions

midpoint	Events observed	Events expected	Sum of ranks
42	23	37.51	-9.5843937
63	17	23.78	-4.9631422
70.5	26	25.25	-.95336946
84.5	36	15.45	15.500905
Total	102	102.00	0

```
            chi2(3) =     35.31
            Pr>chi2 =    0.0000
```

Test for trend of survivor functions

```
            chi2(1) =     22.77
            Pr>chi2 =    0.0000
```

(iii) Comparison of middle two age groups versus the youngest and oldest age group

Test results indicate that the survivorship experience for the middle two age groups is not different from the youngest and the two middle age groups. This difference is evident in the figure above.

```
. gen newage=agegroup
(321 missing values generated)

. recode  newage 1=1 4=. 2=2 3=2
(78 changes made)
```

```
. sts test newage   if yrgrp==1

        failure _d:  fstat
  analysis time _t:  lenfoll

Log-rank test for equality of survivor functions

        |   Events         Events
newage  | observed       expected
--------+-------------------------
1       |       23          28.78
2       |       43          37.22
--------+-------------------------
Total   |       66          66.00

             chi2(1) =      2.07
             Pr>chi2 =    0.1507
```

```
. sts test newage   if yrgrp==1, wilcoxon

        failure _d:  fstat
  analysis time _t:  lenfoll

Wilcoxon (Breslow) test for equality of survivor functions

        |   Events         Events        Sum of
newage  | observed       expected         ranks
--------+----------------------------------------
1       |       23          28.78          -426
2       |       43          37.22           426
--------+----------------------------------------
Total   |       66          66.00             0

             chi2(1) =      1.40
             Pr>chi2 =    0.2362
```

```
. sts test newage   if yrgrp==1, peto

        failure _d:  fstat
  analysis time _t:  lenfoll

Peto-Peto test for equality of survivor functions

        |   Events      Events      Sum of
newage  | observed    expected       ranks
--------+------------------------------------
1       |       23       28.78   -3.5713153
2       |       43       37.22    3.5713153
--------+------------------------------------
Total   |       66       66.00            0

            chi2(1) =      1.45
            Pr>chi2 =    0.2286
```

Next we report the results for comparing the middle two age groups to the oldest age group in grouped cohort 1. The results support a significant difference in the survival experience.

```
. recode  newage 1=.   .= 1 2=2  3=2
(411 changes made)

. sts test newage    if yrgrp==1

        failure _d:  fstat
  analysis time _t:  lenfoll

Log-rank test for equality of survivor functions

        |   Events      Events
newage  | observed    expected
--------+-------------------------
1       |       36       18.79
2       |       43       60.21
--------+-------------------------
Total   |       79       79.00

            chi2(1) =     21.50
            Pr>chi2 =    0.0000
```

```
. sts test newage   if yrgrp==1, wilcoxon

        failure _d:  fstat
   analysis time _t: lenfoll

Wilcoxon (Breslow) test for equality of survivor functions

         |   Events         Events        Sum of
newage   | observed       expected         ranks
---------+------------------------------------------
1        |       36          18.79          1399
2        |       43          60.21         -1399
---------+------------------------------------------
Total    |       79          79.00             0

              chi2(1) =       22.53
              Pr>chi2 =      0.0000
```

```
. sts test newage   if yrgrp==1, peto

        failure _d:  fstat
   analysis time _t: lenfoll

Peto-Peto test for equality of survivor functions

         |   Events         Events        Sum of
newage   | observed       expected         ranks
---------+------------------------------------------
1        |       36          18.79       12.400793
2        |       43          60.21      -12.400793
---------+------------------------------------------
Total    |       79          79.00             0

              chi2(1) =       22.35
              Pr>chi2 =      0.0000
```

7. *For the purposes of this problem, restrict analyses to WHAS data from grouped cohort 1(1975 and 1978). Prepare a table of descriptive statistics for survival time (length of follow-up) for each of the patient characteristic variables in Table 1.4. For age use the four groups in problem 6 above, and for CPK use two groups defined by the median.*

Before obtaining descriptive statistics we converted days of follow-up into years by dividing by 365.25.

Table: Descriptive Statistics of Survival Time (Years) for Subjects in the Worcester Heart Attack Study, 1975 – 1979.

Variable	Category	Frequency	Number of Deaths	Median Survival Time(95 % CI)	Log-Rank Test p-value
Age	42-60	50	23	*	
	61-65	32	17	11.4 (5.21, *)	< 0.001
	66-75	38	26	7.02 (3.95, 9.31)	
	>75	40	36	0.16 (0.02, 1.47)	
CPK	<= 399	80	54	5.61 (2.01, 8.59)	0.304
	> 399	80	48	8.41 (4.53,*)	
Sex	Male	101	62	7.02 (4.64, 11.24)	0.298
	Female	59	40	5.99 (0.94, 8.98)	
SHO	No	152	94	8.0 (5.21, 9.42)	<0.001
	Yes	8	8	0.005 (*, 0.008)	
CHF	No	94	53	9.20 (5.99, *)	0.002
	Yes	66	49	2.01 (0.39, 5.21)	
MIORD	First	96	58	8 (4.53, 12.33)	0.309
	Recurrent	64	44	5.21 (1.47, 8.50)	
MITYPE	Q-wave	110	72	5.12 (1.5, 8.04)	0.133
	Non Q-wave	50	30	8.98 (5.79, *)	

8. *Expand the analyses in problem 7 to include estimates from all 3 cohort groups combined. Note that in this problem the final age interval should be [76,99].*

Solution is a table identical to the one shown above; but with estimates computed from all 3 cohorts.

Chapter Three – Solutions

1. Using the data from the WHAS for grouped cohort 1 (1975 and 1978), with length of follow-up as the survival time variable and status at last follow-up as the censoring variable, do the following:
   (a) Fit the proportional hazards model containing age, sex, peak cardiac enzymes, left heart failure complications and MI order.

```
. stcox age sex cpk chf miord if yrgrp==1, nohr

         failure _d:  fstal
   analysis time _t:  lenfoll

Cox regression -- Breslow method for ties

No. of subjects =         160                Number of obs   =        160
No. of failures =         102
Time at risk    =      415690
                                             LR chi2(5)      =      25.19
Log likelihood  =  -462.44709                Prob > chi2     =     0.0001

------------------------------------------------------------------------------
    _t |
    _d |      Coef.   Std. Err.       z     P>|z|     [95% Conf. Interval]
-------+----------------------------------------------------------------------
   age |   .0339529   .0096279     3.527    0.000     .0150826    .0528231
   sex |   .0112686   .2115987     0.053    0.958    -.4034572    .4259943
   cpk |  -.0000268   .0001735    -0.155    0.877    -.0003669    .0003132
   chf |   .3780826     .21211     1.782    0.075    -.0376454    .7938106
 miord |   .1927073   .2037969     0.946    0.344    -.2067273     .592142
------------------------------------------------------------------------------
```

   (b) Assess the significance of the model using the partial log likelihood ratio test. If it is possible in the software package assess for the significance of the model using the score and Wald tests. Is the statistical decision the same for the three tests?

(i). Null hypothesis: All five coefficients are simultaneously equal to zero

(ii). The partial likelihood ratio test, denoted G, is calculated as follows.
   Log partial likelihood with covariates = − 462.44709
   Log partial likelihood without covariates = −475.04324
   $G = -2(L_p(0) - L_p(\hat{\beta})) = -2(-475.04324 - (-462.44709)) = 25.1923$

   From the Stata output
   LR chi2(5)   =   25.19
   Prob > chi2   =   0.0001

(iii). We reject the null hypothesis and conclude that at least one of the coefficients in the model is significantly related to survival time. The SAS system performs all three tests and we show an example of the output for this problem only. The results are similar for the three tests in other examples in this chapter.

```
The PHREG Procedure

Data Set: A.WHAS1
Dependent Variable: LENFOL
Censoring Variable: FSTAT
Censoring Value(s): 0
Ties Handling: BRESLOW


           Summary of the Number of
           Event and Censored Values

                                 Percent
   Total      Event    Censored  Censored

    160        102        58      36.25


          Testing Global Null Hypothesis: BETA=0

              Without      With
Criterion    Covariates  Covariates    Model Chi-Square

-2 LOG L      950.086     924.894     25.192 with 5 DF (p=0.0001)
Score            .            .        24.838 with 5 DF (p=0.0001)
Wald             .            .        24.719 with 5 DF (p=0.0002)


             Analysis of Maximum Likelihood Estimates

              Parameter    Standard      Wald        Pr >        Risk
Variable  DF  Estimate      Error     Chi-Square  Chi-Square    Ratio

AGE        1    0.033953    0.00963    12.43636     0.0004      1.035
SEX        1    0.011269    0.21160     0.00284     0.9575      1.011
CPK        1   -0.000026818 0.0001735   0.02389     0.8772      1.000
HF         1    0.378083    0.21211     3.17725     0.0747      1.459
MIORD      1    0.192707    0.20380     0.89413     0.3444      1.213
```

(c) *Using the univariate Wald tests, which variables appear not to contribute to the model? Fit a reduced model and test for the significance of the variables removed using the partial log likelihood ratio test.*

The *p* - values for sex, cpk and miord are less than the 0.10 level and they were excluded from the model.

```
. stcox age  chf if yrgrp==1 , nohr

        failure _d:  fstal
   analysis time _t: lenfoll


Cox regression -- Breslow method for ties

No. of subjects =          160                Number of obs   =        160
No. of failures =          102
Time at risk    =       415690
                                              LR chi2(2)      =      24.21
Log likelihood  =   -462.93798                Prob > chi2     =     0.0000


   _t |
   _d |      Coef.    Std. Err.       z     P>|z|     [95% Conf. Interval]
--------+----------------------------------------------------------------
   age |   .034531     .0090802     3.803   0.000     .016734      .0523279
   chf |  .3621702     .2092535     1.731   0.083    -.0479591     .7722994
```

(i). Null hypothesis
All three coefficients for sex, cpk, and miord are simultaneously equal to zero

(ii). The partial likelihood ratio test, denoted G, is calculated as follows.
Full model: Log likelihood = –462.4470
Reduced model: Log likelihood = –462.9379
$G = -2(-462.93798 - (-462.44709)) = 0.98$, df = 3, $p = 0.81$

(iii). We fail to reject the null hypothesis and conclude that sex, cpk, and miord do not contribute to the model significantly.

34 CHAPTER 3 SOLUTIONS

(d) Fit the reduced model in problem 1(c) using Breslow, Efron and exact methods for tied survival times. Compare the estimates of the coefficients and standard errors obtained from the three methods for handling tied survival times. Are the results similar or different?

The coefficients and standard errors obtained form the three methods are similar.

	age		Chf	
	coeff.	std. err	coeff	std. err
Efron	.0346719	.00908	.3636971	.2092176
Breslow	.034531	.0090802	.3621702	.2092535
Exact Marginal Likelihood	.034673	.0091080	.3637089	.2092208

(e) Estimate the baseline survivorship function for the model fit in problem 1(c). Graph the estimated baseline survivorship function versus survival time. What covariate pattern is the "baseline" subject for the fitted model?

```
.sort lenfol
.graph basesv lenfol if yrgrp==1, s(.) c(l) ylab xlab yscale(0,1)
```

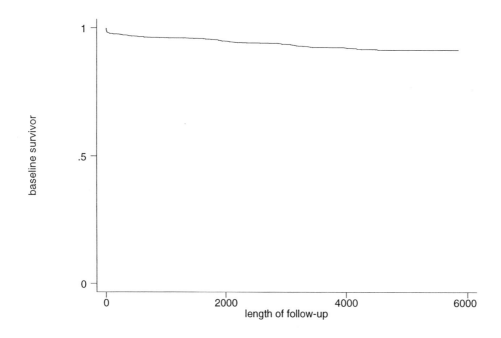

Baseline subject: age=0, no left heart failure complications

(f) Repeat problem 1(e) using age centered at the median age of 65 years. Explain why the range of the estimated survivorship functions in problems 1(e) and 1(f) are different.

```
. drop basesv
. gen age65=age-65
. stcox age65 chf if yrgrp==1, nohr basesurv(basesv)

. graph basesv lenfol, s(.) c(l) ylab xlab yscale(0,1)
```

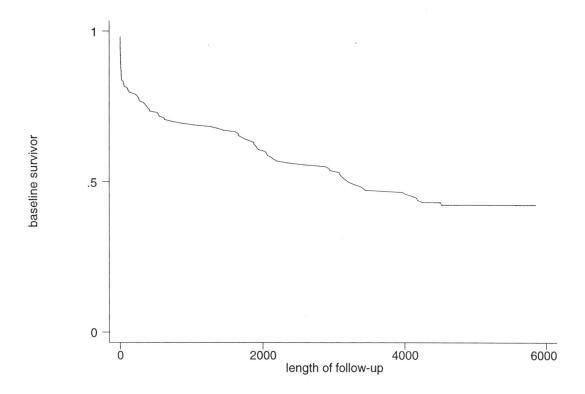

Baseline subject: age=65, no left heart failure complications

(g) Using the model fit in problem 1(f) estimate the value of the survivorship function for each subject at his or her respective observed value of time. Graph the values of the estimated survivorship function versus survival time. Why is there scatter in this plot that was not present in the graphs in problems 1(e) and 1(f)?

```
. predict xb, xb
. gen surv=basesv^exp(xb)
 (321 missing values generated)
. sort lenfol
```

```
. graph surv lenfol, s(o) ylab xlab yscale(0,1)
```

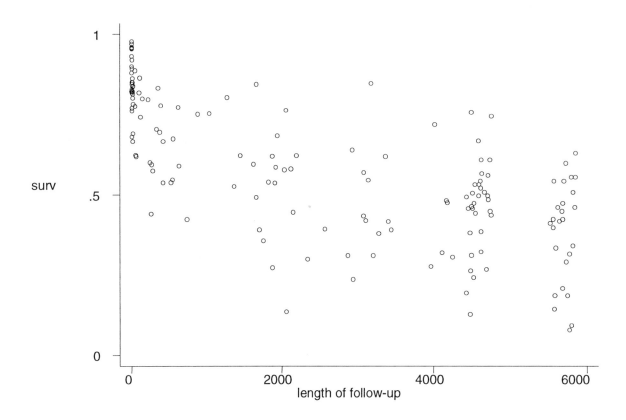

Scatter is due to the fact that the function is evaluated at each individual's unique set of covariate

2. *Repeat problem 1 for each of the other grouped cohorts*

The results are similar to problem 1 for each of the three grouped cohorts and are not shown.

3. *Repeat problem 1 using all the data from the WHAS (i.e., ignore cohort).*

Here we show the output as the reduced model is larger than that used in the previous problems.

(a)   Fit the proportional hazards model containing age, sex, peak cardiac enzymes, left heart failure complications and MI order.

```
. stcox age sex cpk chf miord, nohr

        failure _d:  fstat
   analysis time _t: lenfol

Cox regression -- Breslow method for ties

No. of subjects =          481                  Number of obs   =         481
No. of failures =          249
Time at risk    =       834555
                                                LR chi2(5)      =      118.21
Log likelihood  =   -1361.5039                  Prob > chi2     =      0.0000

------------------------------------------------------------------------------
    _t           |
    _d           |    Coef.   Std. Err.      z    P>|z|    [95% Conf. Interval]
-----------------+------------------------------------------------------------
         age     |  .0353064   .0059153    5.969   0.000    .0237126    .0469003
         sex     |  .0830084    .133492    0.622   0.534   -.1786311    .3446479
         cpk     |  .0001498   .0000617    2.429   0.015    .0000289    .0002706
         chf     |    .7679   .1361474    5.640   0.000     .501056    1.034744
       miord     |  .398466   .1296629    3.073   0.002    .1443313    .6526006
------------------------------------------------------------------------------
```

(b)   Assess the significance of the model using the partial log likelihood ratio test.

(i). Null hypothesis: All five coefficients are simultaneously equal to zero

(ii). The partial likelihood ratio test, denoted G, is calculated as follows.
Log partial likelihood with covariates = -1361.5039
Log partial likelihood without covariates = -1420.6083

$$G = -2(L_p(0) - L_p(\hat{\beta})) = -2(-1420.6083 - (-1361.5039)) = 118.2088$$

From Stata output
```
LR chi2(5)   = 118.21
Prob > chi2  = 0.0000
```

(iii). We reject the null hypothesis and conclude that at least one of the coefficients in the model is significantly related to survival time.

(c) Using the univariate Wald tests, which variables appear not to contribute to the model? Fit a reduced model and test for the significance of the variables removed using the partial log likelihood ratio test.

Only the coefficient for sex is not significant so we drop it from the model.

```
. stcox age cpk chf miord, nohr                sex is dropped

        failure _d:  fstat
   analysis time _t: lenfol

Cox regression -- Breslow method for ties

No. of subjects =         481                  Number of obs   =        481
No. of failures =         249
Time at risk    =      834555
                                               LR chi2(4)      =     117.82
Log likelihood  =   -1361.6969                 Prob > chi2     =     0.0000

     _t |
     _d |     Coef.   Std. Err.      z    P>|z|     [95% Conf. Interval]
--------+----------------------------------------------------------------
    age |  .0362813   .0057068    6.358   0.000     .0250962    .0474665
    cpk |  .0001473   .0000616    2.392   0.017     .0000266     .000268
    chf |  .7689733   .1360253    5.653   0.000     .5023685    1.035578
  miord |  .4006601   .1296258    3.091   0.002     .1465982     .654722
```

(i) Null hypothesis
   The coefficient for sex is equal to zero.

(ii) The partial likelihood ratio test, denoted G, is calculated as follows.
Full model: Log likelihood = -1361.5039
Reduced model: Log likelihood = -1361.6969
   $G = -2(-1361.6969 - (-1361.5039)) = 0.386$, df = 1, $p = 0.5344$

(iii) We fail to reject the null hypothesis and conclude that sex does not contribute model significantly.

(d) Fit the reduced model in problem (c) using Breslow, Efron and exact methods for tied survival times. Compare the estimates of the coefficients and standard errors obtained from the three methods for handling tied survival times. Are the results similar or different?

	age		cpk		chf		miord	
	coeff.	st. err	coeff.	st. err	coeff	st. err	Coeff.	st. err
Efron	.036401	.00571	.00015	.00006	.7715	.1360	.4029	.12962
Breslow	.036281	.00571	.00015	.00006	.7690	.1360	.4007	.12963
Exact Marginal Likelihood	.036403	.00571	.00015	.00006	.7715	.1360	.4029	.12962

The coefficients and standard errors obtained form the three methods are similar

(e) Estimate the baseline survivorship function for the model fit in problem (c). Graph the estimated baseline survivorship function versus survival time. What covariate pattern is the "baseline" subject for the fitted model?

```
. stcox age cpk chf miord, nohr basesurv(basesv)
. sort lenfol
```

```
. graph basesv lenfol, s(.) c(l) ylab xlab yscale(0,1)
```

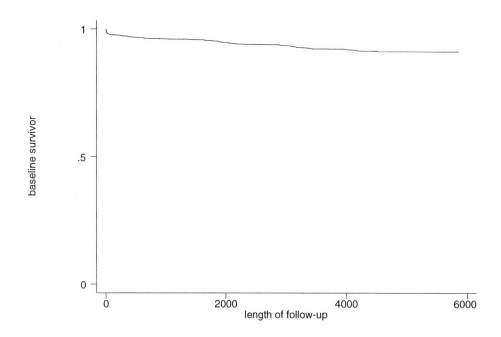

Baseline subject: age=0, no left heart failure complications, peak cardiac enzyme level=0, first MI disorder

(f) *Repeat problem (e) using age centered at the median age of 65 years. Explain why the range of the estimated survivorship functions in problems 1(e) and 1(f) are different.*

```
. drop basesv
. gen age65=age-65
. stcox age65 cpk chf miord, nohr basesurv(basesv)
. sort lenfol
```

```
. graph basesv lenfol, s(.) c(l) ylab xlab yscale(0,1)
```

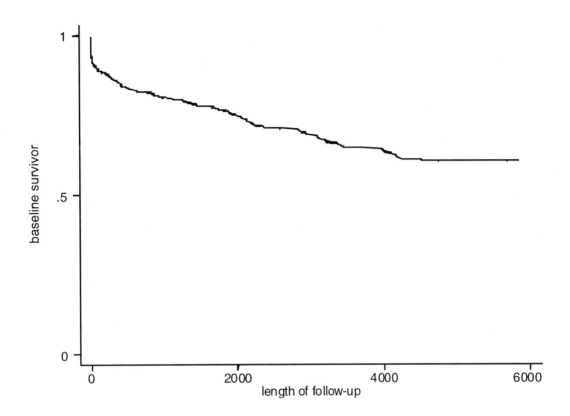

Baseline subject: age = 65, no left heart failure complications, peak cardiac enzyme level = 0, first MI disorder

> (g) Using the model fit in problem (e) estimate the value of the survivorship function for each subject at his or her respective observed value of time. Graph the values of the estimated survivorship function versus survival time. Why is the range of the scatter in this plot that was not present in the graphs in problems (e) and (f)?

```
. predict xb, xb
. gen surv=basesv^exp(xb)
. sort lenfol
```

```
. graph surv lenfol, s(o) ylab xlab yscale(0,1)
```

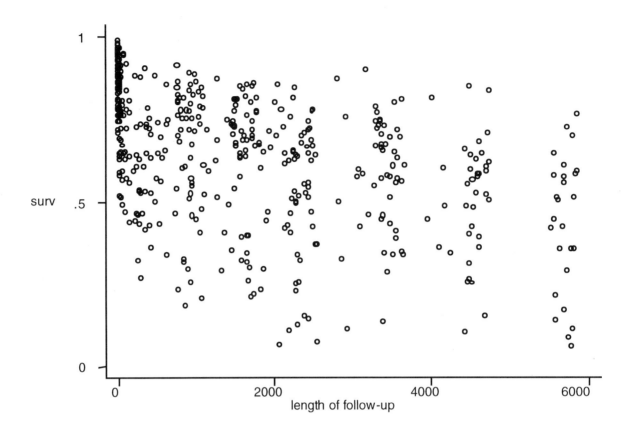

As before, the range is different due to the distinct values of the covariates used to compute the individual survival probabilities.

Chapter Four – Solutions

1. *Using all the data from the WHAS (i.e., ignore cohort), with length of follow-up as the survival time variable and status at last follow-up as the censoring variable, do the following:*

    (a) *Fit the proportional hazards model containing sex and estimate the hazard ratio, pointwise and with a 90 percent confidence interval. Interpret the point and interval estimates.*

```
. stcox   sex , level(90) noshow nolog

Cox regression -- Breslow method for ties

No. of subjects =          481                 Number of obs    =        481
No. of failures =          249
Time at risk    =       834555
                                               LR chi2(1)       =       9.13
Log likelihood  =    -1416.0443                Prob > chi2      =     0.0025

------------------------------------------------------------------------------
     _t  |
     _d  | Haz. Ratio   Std. Err.      z    P>|z|    [90% Conf. Interval]
---------+--------------------------------------------------------------------
     sex |   1.472518    .1872124     3.04  0.002     1.194647     1.81502
------------------------------------------------------------------------------
```

Females (sex=1) are dying at a rate that is estimated to be 1.47 times greater than males (sex=0) and the rate could be as low as 1.19 or as high as 1.82 times with 90 percent confidence.

    (b) *Add age to the model fit in 1(a). Is age a confounder of the effect of sex? Explain the reasons for your answer.*

```
. stcox   sex , nohr noshow nolog         crude estimator

Cox regression -- Breslow method for ties

No. of subjects =          481                 Number of obs    =        481
No. of failures =          249
Time at risk    =       834555
                                               LR chi2(1)       =       9.13
Log likelihood  =    -1416.0443                Prob > chi2      =     0.0025

------------------------------------------------------------------------------
     _t  |
     _d  |     Coef.    Std. Err.      z    P>|z|    [95% Conf. Interval]
---------+--------------------------------------------------------------------
     sex |   .3869737    .1271376     3.04  0.002     .1377886    .6361588
------------------------------------------------------------------------------
```

```
. stcox   sex age , nohr noshow nolog             age adjusted estimator

Cox regression -- Breslow method for ties

No. of subjects =         481              Number of obs    =        481
No. of failures =         249
Time at risk    =      834555
                                           LR chi2(2)       =      70.47
Log likelihood  =   -1385.3756             Prob > chi2      =     0.0000

------------------------------------------------------------------------------
       _t |
       _d |      Coef.   Std. Err.      z     P>|z|     [95% Conf. Interval]
----------+-------------------------------------------------------------------
      sex |   .0840387   .1326917     0.63    0.527    -.1760323    .3441097
      age |   .0428161   .0055325     7.74    0.000     .0319725    .0536596
------------------------------------------------------------------------------
```

Age is a confounder as the percent change in the coefficient for sex is

$$\Delta\beta_{sex}\% = 100 \times \frac{0.39 - 0.08}{0.08} = 388\%$$

(c) *Is there a significant interaction between age and sex? (Use alpha=0.1 for this problem).*

```
. gen agexsex=age*sex

. stcox sex age   agexsex, nohr

          failure _d:  fstat
   analysis time _t:  lenfol

Iteration 0:   log likelihood = -1420.6083
Iteration 1:   log likelihood = -1384.5071
Iteration 2:   log likelihood = -1383.6328
Iteration 3:   log likelihood = -1383.6325
Refining estimates:
Iteration 0:   log likelihood = -1383.6325

Cox regression -- Breslow method for ties

No. of subjects =         481              Number of obs    =        481
No. of failures =         249
Time at risk    =      834555
                                           LR chi2(3)       =      73.95
Log likelihood  =   -1383.6325             Prob > chi2      =     0.0000

------------------------------------------------------------------------------
       _t |
       _d |      Coef.   Std. Err.      z     P>|z|     [95% Conf. Interval]
----------+-------------------------------------------------------------------
      sex |   1.552902   .7923099     1.96    0.050     3.44e-06    3.105801
      age |   .0517544   .0073028     7.09    0.000     .0374411    .0660676
  agexsex |  -.0204025   .0108827    -1.87    0.061    -.0417322    .0009271
------------------------------------------------------------------------------
```

The *p*-value for the Wald statistic is 0.061. Thus at the 0.1 level we conclude that there is a significant interaction.

> *(d)   Using the model fit in 1(c) estimate the hazard ratio, pointwise and 90 percent confidence interval, for gender at age 50, 60, 65, 70 and 80.*

Age	HR	90% CI
50	1.70	1.09, 2.66
60	1.39	1.03, 1.87
65	1.25	.98, 1.6
70	1.13	.91, 1.40
80	.92	.71, 1.20

Up to age 80 the estimated hazard ratios are all greater than 1 but decrease with increasing age. Confidence intervals support a significant sex effect for subjects 60 and younger. We conclude that females are dying at a significantly higher rate than males for age 60 and younger.

Results from using lincom to obtain the HR's and CI's is as shown below.

```
. lincom sex + 50*agexsex, hr level(90)

 ( 1)   sex + 50.0 agexsex = 0.0

------------------------------------------------------------------------------
     _t | Haz. Ratio   Std. Err.      z    P>|z|     [90% Conf. Interval]
--------+---------------------------------------------------------------------
    (1) |   1.703653    .4607164     1.97   0.049     1.091944    2.658043
------------------------------------------------------------------------------
```

```
. lincom sex + 60*agexsex, hr level(90)

 ( 1)   sex + 60.0 agexsex = 0.0

------------------------------------------------------------------------------
     _t | Haz. Ratio   Std. Err.      z    P>|z|     [90% Conf. Interval]
--------+---------------------------------------------------------------------
    (1) |    1.38923    .2533839     1.80   0.071     1.029159    1.875278
------------------------------------------------------------------------------
```

```
. lincom sex + 65*agexsex, hr level(90)

 ( 1)   sex + 65.0 agexsex = 0.0

------------------------------------------------------------------------------
     _t | Haz. Ratio   Std. Err.      z    P>|z|     [90% Conf. Interval]
--------+---------------------------------------------------------------------
    (1) |     1.2545    .1869868     1.52   0.128     .9817355    1.603048
------------------------------------------------------------------------------
```

```
. lincom sex +70*agexsex, hr level(90)

 ( 1)  sex + 70.0 agexsex = 0.0

------------------------------------------------------------------------------
         _t | Haz. Ratio   Std. Err.      z    P>|z|    [90% Conf. Interval]
-------------+----------------------------------------------------------------
        (1) |   1.132836    .1480875     0.95   0.340     .9136608    1.404588
------------------------------------------------------------------------------
```

```
. lincom sex +80*agexsex, hr level(90)

 ( 1)  sex + 80.0 agexsex = 0.0

------------------------------------------------------------------------------
         _t | Haz. Ratio   Std. Err.      z    P>|z|    [90% Conf. Interval]
-------------+----------------------------------------------------------------
        (1) |   .9237614     .144877    -0.51   0.613     .7137161    1.195623
------------------------------------------------------------------------------
```

(e)   Using the model fit in 1(c), estimate (pointwise and with a 90 percent confidence interval) the hazard ratio for a 10-year increase in age for each gender.

Gender	HR	90% CI
Male	1.68	1.49, 1.89
Female	1.37	1.19, 1.56

We conclude that the effect of increasing age is significant for both males and females with a slightly stronger effect among males.

```
. lincom 10*agexsex+ 10*age, hr level(90)         Female

 ( 1)  10.0 age + 10.0 agexsex = 0.0

------------------------------------------------------------------------------
         _t | Haz. Ratio   Std. Err.      z    P>|z|    [90% Conf. Interval]
-------------+----------------------------------------------------------------
        (1) |    1.36823    .1117619     3.84   0.000     1.196213    1.564984
------------------------------------------------------------------------------
```

```
. lincom 10*age, hr level(90)

 ( 1)  10.0 age = 0.0

------------------------------------------------------------------------------
         _t | Haz. Ratio   Std. Err.      z    P>|z|    [90% Conf. Interval]
-------------+----------------------------------------------------------------
        (1) |   1.677901     .122534     7.09   0.000     1.487985    1.892056
------------------------------------------------------------------------------
```

(f) *Using the model fit in 1(c), compute, and then graph, the estimated survivorship functions for 65-year-old males and females. Interpret the survivorship experience presented in this graph.*

```
. stcox sex age agesex, bases(basesrv) noshow nolog nohr

Cox regression -- Breslow method for ties

No. of subjects =          481                Number of obs   =       481
No. of failures =          249
Time at risk    =   2284.887068
                                              LR chi2(3)      =     73.95
Log likelihood  =   -1383.6325                Prob > chi2     =    0.0000

------------------------------------------------------------------------------
       _t |
       _d |      Coef.   Std. Err.      z    P>|z|     [95% Conf. Interval]
----------+-------------------------------------------------------------------
      sex |   1.552902   .7923099     1.96   0.050     3.44e-06    3.105801
      age |   .0517544   .0073028     7.09   0.000     .0374411    .0660676
   agesex |  -.0204025   .0108827    -1.87   0.061    -.0417322    .0009271
```

. gen survm=basesrv^exp(_b[sex]*0+_b[age]*65+_b[agesex]*65*0) if sex==0
(194 missing values generated)

. gen survf=basesrv^exp(_b[sex]*1+_b[age]*65+_b[agesex]*65*1) if sex==1
(287 missing values generated)

```
. list survm lenfol if survm <=0.52 & survm >=0.48 & fstat==1

           survm      lenfoll
   371.  .5155363      3139
   372.  .5109465      3171
   382.  .4968842      3361
```

```
. list survf lenfol if survf <=0.52 & survf >=0.48 & fstat==1

           survf      lenfoll
   310.  .5136493      2131
   311.  .5101687      2136
   313.  .5031864      2177
   314.  .4996922      2187
   318.  .4961397      2217
   319.  .4925629      2238
   336.  .4884717      2335
```

. sort lenfol

48 CHAPTER 4 SOLUTIONS

```
graph survm survf lenfol, s(..) c(JJ)   yscale(0,1) /*
*/ ylabel(.1,.2,.3,.4,.5,.6,.7,.8,.9,1) yline(0.5) /*
*/ xlabel(500,1000,1500,2000,2500,3000,3500,4000,4500,5000) xline(2187,3361)
```

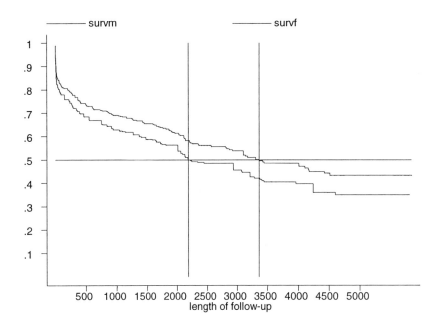

(g) Using the graph in 1(f), estimate the median survival time for 65-year-old males and females.

From the list we see that the estimated median survival times for females is 2187 and males is 3361. These lines have been added to the graph.

2. *Repeat problem 1, parts (b)-(d), with age broken into four groups at its quartiles. In part (d) estimate hazard ratios for each age group.*

   (b) Is age a confounder of the effect of sex? Explain the reasons for your answer.

```
. sum age, detail
                    patient age (per chart)
-------------------------------------------------------------
            Percentiles      Smallest
 1%             35              24
 5%             47              28
10%             52              33        Obs                481
25%             59              34        Sum of Wgt.        481
50%             68                        Mean           67.48441
                              Largest     Std. Dev.      12.68054
75%             77              95
90%             84              96        Variance       160.7961
95%             88              97        Skewness      -.1882374
99%             94              98        Kurtosis       2.893568
```

```
. gen ageqtl = age
. recode ageqtl  24/59=1 60/68=2 69/77=3 78/98=4
(481 changes made)
```

```
. stcox sex, nolog noshow nohr

Cox regression -- Breslow method for ties

No. of subjects =         481              Number of obs   =      481
No. of failures =         249
Time at risk    =  2284.887068
                                           LR chi2(1)      =     9.13
Log likelihood  =   -1416.0443             Prob > chi2     =   0.0025

------------------------------------------------------------------------
     _t |
     _d |      Coef.   Std. Err.      z    P>|z|    [95% Conf. Interval]
--------+---------------------------------------------------------------
    sex |   .3869737   .1271376    3.04   0.002    .1377886    .6361588
------------------------------------------------------------------------
```

```
. xi:stcox sex i.ageqtl, nolog noshow nohr
i.ageqtl          _Iageqtl_1-4      (naturally coded; _Iageqtl_1 omitted)

Cox regression -- Breslow method for ties

No. of subjects =         481              Number of obs   =      481
No. of failures =         249
Time at risk    =  2284.887068
                                           LR chi2(4)      =    79.28
Log likelihood  =   -1380.9683             Prob > chi2     =   0.0000

------------------------------------------------------------------------
      _t |
      _d |      Coef.   Std. Err.      z    P>|z|    [95% Conf. Interval]
---------+--------------------------------------------------------------
     sex |   .0128192   .1363285    0.09   0.925   -.2543798    .2800183
_Iageqtl_2|  .5261945   .2144111    2.45   0.014    .1059566    .9464324
_Iageqtl_3| 1.090515    .202849     5.38   0.000    .6929382   1.488092
_Iageqtl_4| 1.554501    .2078837    7.48   0.000   1.147056    1.961946
------------------------------------------------------------------------
```

Age is clearly a confounder as the change in the coefficient for sex is 2919%, age adjustment is needed.

$$\Delta\beta_{sex}\% = 100 \times \frac{0.38697 - 0.012819}{0.012819} = 2919\%$$

(c)   Is there a significant interaction between age and sex? (Use alpha=0.1 for this problem).

```
. xi:stcox i.sex*i.ageqtl, nolog noshow nohr
i.sex             _Isex_0-1        (naturally coded; _Isex_0 omitted)
i.ageqtl          _Iageqtl_1-4     (naturally coded; _Iageqtl_1 omitted)
i.sex*i.ageqtl    _IsexXage_#_#    (coded as above)

Cox regression -- Breslow method for ties

No. of subjects  =          481              Number of obs   =        481
No. of failures  =          249
Time at risk     =   2284.887068
                                             LR chi2(7)      =      89.20
Log likelihood   =   -1376.0083              Prob > chi2     =     0.0000

------------------------------------------------------------------------------
      _t |
      _d |     Coef.   Std. Err.      z     P>|z|     [95% Conf. Interval]
---------+--------------------------------------------------------------------
  _Isex_1 |   .2866275   .3685195    0.78   0.437    -.4356576    1.008912
_Iageqtl_2|   .5177572   .2608159    1.99   0.047     .0065673    1.028947
_Iageqtl_3|   .9915541   .2516646    3.94   0.000     .4983006    1.484808
_Iageqtl_4|   1.983629   .2542433    7.80   0.000     1.485322    2.481937
_IsexXage_~2| -.0780016   .4644459  -0.17   0.867    -.9882988    .8322956
_IsexXage_~3|  .1020738   .436816    0.23   0.815    -.7540699    .9582175
_IsexXage_~4| -.8365369   .4290245  -1.95   0.051    -1.677409    .0043356

. lrtest, saving(0)
```

```
. quietly xi:stcox sex i.ageqtl, nolog noshow nohr
. lrtest
       Cox:   likelihood-ratio test                  chi2(3)   =       9.92
                                                     Prob > chi2 =     0.0193
```

Thus we consider age as an effect modifier as the likelihood ratio test is significant. The *p*-values for the Wald statistics for the interaction coefficients suggests that the interaction may be most important for the fourth quartile of age.

(d)   Using the model fit in 2(c) estimate the hazard ratios for sex within each age group.

```
. quietly xi:stcox i.sex*i.ageqtl, nolog noshow nohr
. lincom _b[_Isex_1], hr level(90)

 ( 1)  _Isex_1 = 0.0

------------------------------------------------------------------------------
      _t | Haz. Ratio   Std. Err.      z     P>|z|     [90% Conf. Interval]
---------+--------------------------------------------------------------------
     (1) |   1.331928   .4908415    0.78   0.437     .7264881    2.441929
------------------------------------------------------------------------------
```

```
. lincom  _b[ _Isex_1] + _b[ _IsexXage_1_2], hr level(90)

 ( 1)  _Isex_1 + _IsexXage_1_2 = 0.0

------------------------------------------------------------------------------
          _t |  Haz. Ratio   Std. Err.      z    P>|z|     [90% Conf. Interval]
-------------+----------------------------------------------------------------
         (1) |   1.231984    .3482462     0.74   0.460     .7738891    1.961243
------------------------------------------------------------------------------
```

```
. lincom  _b[ _Isex_1] + _b[ _IsexXage_1_3], hr level(90)

 ( 1)  _Isex_1 + _IsexXage_1_3 = 0.0

------------------------------------------------------------------------------
          _t |  Haz. Ratio   Std. Err.      z    P>|z|     [90% Conf. Interval]
-------------+----------------------------------------------------------------
         (1) |   1.475064     .345587     1.66   0.097     1.00334     2.168569
------------------------------------------------------------------------------
```

```
. lincom  _b[ _Isex_1] + _b[ _IsexXage_1_4], hr level(90)

 ( 1)  _Isex_1 + _IsexXage_1_4 = 0.0

------------------------------------------------------------------------------
          _t |  Haz. Ratio   Std. Err.      z    P>|z|     [90% Conf. Interval]
-------------+----------------------------------------------------------------
         (1) |   .5770021    .1263596    -2.51   0.012     .4024754     .8272094
------------------------------------------------------------------------------
```

Age Quartile	HR	90% CI
1	1.33	0.72, 2.44
2	1.23	0.77, 1.96
3	1.48	1.00, 2.17
4	0.58	0.40, 0.83

The results in the above table show that the hazard ratio for females is not significantly different from males in the first two quartiles. In the third quartile females are dying at a rate that is about 1.5 times higher than males and is just significant at the 10 percent level. In the fourth quartile females are dying at a rate that is significantly lower than males, about 42 percent lower. The picture of the age sex interaction is more complicated than that obtained when we treated age as continuous. This difference could be due to a non-linear effect of age itself. This will be explored in later chapters.

3. *Using the data from the WHAS (i.e., ignore cohort), with length of follow-up as the survival time variable and status at last follow-up as the censoring variable, do the following:*

   (a) *Fit the proportional hazards model containing age centered at 65 years, sex, peak cardiac enzymes centered at 650, cardiogenic shock complications, left heart failure complications and MI order and obtain the estimated baseline survivorship function. (Note: In this problem, ignore the possible sex age interaction investigated in problems 1 and 2.) Estimate hazard ratios (via point estimates and 95 percent confidence intervals) for each variable in the model.*

```
. gen age_c=age-65
. gen cpk_c=cpk-650

. stcox age_c sex cpk_c sho chf miord, noshow nolog nohr bases(basesrv)

Cox regression -- Breslow method for ties

No. of subjects =         481                  Number of obs   =        481
No. of failures =         249
Time at risk    =   2284.887068
                                               LR chi2(6)      =     173.76
Log likelihood  =   -1333.7304                 Prob > chi2     =     0.0000

------------------------------------------------------------------------------
        _t   |
        _d   |     Coef.   Std. Err.      z    P>|z|     [95% Conf. Interval]
-------------+----------------------------------------------------------------
     age_c   |  .0344091   .0059422     5.79   0.000     .0227626    .0460556
       sex   |  .0144279    .133479     0.11   0.914    -.2471861    .2760418
     cpk_c   |  .0000877   .0000639     1.37   0.170    -.0000376    .0002129
       sho   |  1.760383   .2099816     8.38   0.000     1.348827    2.17194
       chf   |  .5885312   .1428963     4.12   0.000     .3084596    .8686028
    miord1   |  .2837548   .1325252     2.14   0.032     .0240101    .5434995
------------------------------------------------------------------------------
```

```
. stcox age_c sex cpk_c sho chf miord

Cox regression -- Breslow method for ties

No. of subjects =         481                  Number of obs   =        481
No. of failures =         249
Time at risk    =    2284.887068
                                               LR chi2(6)      =     173.76
Log likelihood  =    -1333.7304                Prob > chi2     =     0.0000

------------------------------------------------------------------------------
          _t |
          _d | Haz. Ratio   Std. Err.      z    P>|z|     [95% Conf. Interval]
-------------+----------------------------------------------------------------
       age_c |   1.035008    .0061502     5.79   0.000     1.023024    1.047133
         sex |   1.014532    .1354187     0.11   0.914      .7809953   1.317903
       cpk_c |   1.000088    .0000639     1.37   0.170      .9999624   1.000213
         sho |   5.814665    1.220973     8.38   0.000     3.852903    8.775288
         chf |   1.801341    .2574049     4.12   0.000     1.361327    2.383578
      miord1 |   1.328107    .1760077     2.14   0.032     1.024301    1.722023
------------------------------------------------------------------------------
```

The hazard ratios and their 95 percent confidence intervals for sex, sho and chf may be read from the above table. We use lincom, below, to estimate hazard ratios for a 10 year increase in age and for a 1000 point increase in cpk.

```
. lincom 10*_b[age_c],hr

 ( 1)  10.0 age_c = 0.0

------------------------------------------------------------------------------
          _t | Haz. Ratio   Std. Err.      z    P>|z|     [95% Conf. Interval]
-------------+----------------------------------------------------------------
         (1) |   1.410707    .0838273     5.79   0.000     1.255615    1.584956
------------------------------------------------------------------------------
```

```
. lincom 1000*_b[cpk_c],hr

 ( 1)  1000.0 cpk_c = 0.0

------------------------------------------------------------------------------
          _t | Haz. Ratio   Std. Err.      z    P>|z|     [95% Conf. Interval]
-------------+----------------------------------------------------------------
         (1) |   1.091608    .0697522     1.37   0.170      .9631106   1.23725
------------------------------------------------------------------------------
```

54 CHAPTER 4 SOLUTIONS

(b)   Using the methods for the modified risk score, compute and graph the estimated survivorship functions for subjects wit hand without cardiogenic shock complications. Use the estimated survivorship functions to estimate the median survival time.

```
. predict r, xb
. gen rm=r-_b[sho]*sho
```

```
. sum rm, detail

                              rm
-------------------------------------------------------------
      Percentiles      Smallest
 1%    -.855056       -1.373959
 5%    -.5376227      -1.138256
10%    -.3383609      -1.088332      Obs                 481
25%    -.02165        -.980021       Sum of Wgt.         481

50%     .4400124                     Mean            .4587342
                       Largest       Std. Dev.       .6277771
75%     .9504978       1.818627
90%    1.313061        1.838158      Variance        .3941041
95%    1.479501        1.911087      Skewness        .0224109
99%    1.805203        2.00521       Kurtosis        2.405814
```

```
. gen lpred_0=0.440012
. gen lpred_1=0.4400124+_b[sho]*1
. gen S0=basesrv^exp(lpred_0) if sho==0
(38 missing values generated)
. gen S1=basesrv^exp(lpred_1) if sho==1
(443 missing values generated)
```

```
. list S0 lenfol if S0<=0.52 & S0 >=0.48 & fstat==1

              S0       lenfoll
364.    .5183489         2827
365.    .5129465         2868
366.    .5075451         2922
367.    .5021158         2936
368.    .4966474         3071
369.    .4911915         3078
370.    .4857402         3103
371.    .4802936         3139
```

```
. list S1 lenfol if S1<=0.55 & S1 >=0.40 & fstat==1

              S1        lenfoll
  76.    .5281172          15
  77.    .5281172          15
  78.    .5281172          15
  83.    .5051625          19
  92.    .4549218          28
 100.    .4128512          48
```

The above lists show that the covariate adjusted median survival time for subjects with sho = 0 is 3071 days and for subject with sho = 1 is 28 days. These values have been added to the graph below. Note that the graphs are drawn only for subjects in their respective groups. In particular it would be an inappropriate extrapolation of the results for sho = 1 to draw its graph over all subjects.

```
. sort lenfoll

. graph S0 S1 lenfoll, s(..) c(JJ)  ylab(.1,.2,.3,.4,.5,.6,.7,.8,.9,1.0) /*
> */  yscale(0,1) yline(0.5) xline(28, 3071)
```

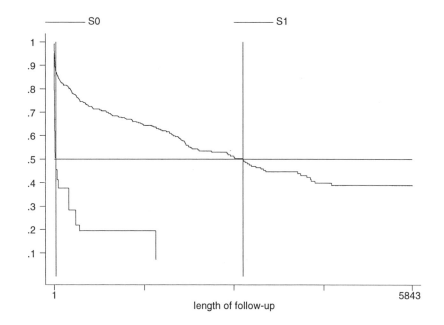

Chapter Five – Solutions

1. An important step in any model building process is assessing the scale of continuous variables in the model. The two continuous variables, AGE and CPK, in the WHAS present a challenge. Use the methods discussed in this chapter to assess the scale of AGE when it is the only covariate in a proportional hazards model. Repeat this process for CPK. In this problem, pay particular attention to the effect that a few subjects with either small or large values of the covariate can have on the methods for assessing the scale of a covariate.

Note: We will assess the scale of age and leave CPK for you to do your own, as the methods are the same.

(i). Quartile design variable method

To check the linearity of the grouped age variable, we graph the coefficients versus the age group midpoint.

```
. sum age, detail
                patient age (per chart)
-------------------------------------------------------------
      Percentiles      Smallest
 1%        35             24
 5%        47             28
10%        52             33          Obs                 481
25%        59             34          Sum of Wgt.         481
50%        68                         Mean           67.48441
                      Largest         Std. Dev.      12.68054
75%        77             95
90%        84             96          Variance       160.7961
95%        88             97          Skewness      -.1882374
99%        94             98          Kurtosis       2.893568
```

. **capture drop ageqtl**

. **gen ageqtl = age**

. **recode ageqtl   24/59=1 60/68=2 69/77=3 78/98=4**
(481 changes made)

```
. xi:stcox i.ageqtl , nohr noshow nolog
i.ageqtl          _Iageqtl_1-4        (naturally coded; _Iageqtl_1 omitted)

Cox regression -- Breslow method for ties

No. of subjects =         481                Number of obs   =        481
No. of failures =         249
Time at risk    =    2284.887068
                                             LR chi2(3)      =      79.27
Log likelihood  =   -1380.9727               Prob > chi2     =     0.0000

------------------------------------------------------------------------------
         _t |
         _d |     Coef.   Std. Err.       z    P>|z|     [95% Conf. Interval]
------------+-----------------------------------------------------------------
 _Iageqtl_2 |  .5280596   .2134871     2.47    0.013     .1096325    .9464867
 _Iageqtl_3 |  1.092999   .2011176     5.43    0.000     .6988154    1.487182
 _Iageqtl_4 |   1.56083   .1966838     7.94    0.000     1.175337    1.946324
------------------------------------------------------------------------------

. clear

. input loghaz age_mid

        loghaz       age_mid
  1.  0              41.5
  2.  0.528          64
  3.  1.093          73
  4.  1.56           88
  5.
```

. graph loghaz age_mid, c(l) xlabel(41.5, 64, 73,88)

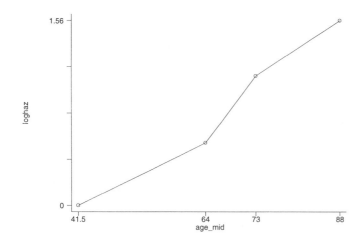

The plot of the coefficients for age support an assumption of linearity in the log hazard.

(ii) Fractional polynomials

We use the method of fractional polynomial to determine whether the data support the hypothesis that the covariate is linear in the log-hazard. If not, this procedure suggests the type of transformation of the covariate. Fractional polynomial functions are similar to conventional polynomials in that they include powers of the covariate, but non-integer and negative powers are also allowed.

```
. use whas
```

```
. fracpoly cox lenfol age, dead(fstat) compare
........
-> gen double Iage__1 = X^-2-.022 if e(sample)
-> gen double Iage__2 = X^-1-.1482 if e(sample)
   (where: X = age/10)

Iteration 0:   log likelihood = -1420.6083
Iteration 1:   log likelihood = -1383.3366
Iteration 2:   log likelihood = -1382.5156
Iteration 3:   log likelihood = -1382.4484
Iteration 4:   log likelihood = -1382.4476
Refining estimates:
Iteration 0:   log likelihood = -1382.4476

Cox regression -- Breslow method for ties
Entry time 0                                    Number of obs   =       481
                                                LR chi2(2)      =     76.32
                                                Prob > chi2     =    0.0000
Log likelihood = -1382.4476                     Pseudo R2       =    0.0269

       lenfol |
        fstat |     Coef.   Std. Err.       z    P>|z|     [95% Conf. Interval]
--------------+----------------------------------------------------------------
      Iage__1 |  92.87998   15.26461     6.08   0.000     62.96188    122.7981
      Iage__2 | -50.03738   6.685254    -7.48   0.000    -63.14023   -36.93452

Deviance: 2764.895. Best powers of age among 44 models fit: -2 -1.

Fractional polynomial model comparisons:
------------------------------------------------------------
age              df    Deviance    Gain    P(term) Powers
------------------------------------------------------------
Not in model     0     2841.217     --       --
Linear           1     2771.152    0.000    0.000   1
m = 1            2     2771.152    0.000    1.000   1
m = 2            4     2764.895    6.256    0.044   -2 -1
------------------------------------------------------------
```

The significance level reported in the 4th column of the second row is for the partial likelihood ratio test of age entering the model as a linear term. The best power when age enters the model with a single term is 1.

The best power when age enters the model with a two terms is -2 and -1. The interpretation is that the two terms are $1/(age)^2$ and $1/age$. The approximate partial likelihood ratio test comparing the use of two terms versus the best one term model is G = 2771.152 - 2764.895 = 6.256, which is reported in the Gain column. The p-value for the test is with 2 degree of freedom is $\Pr(\chi^2(2) \geq 6.256) = 0.044$. Thus we conclude that the two term model ($m = 2$) model may fit better than one term ($m = 1$). To examine this further we look at a plot of the $m = 2$ function.

. `fracplot`

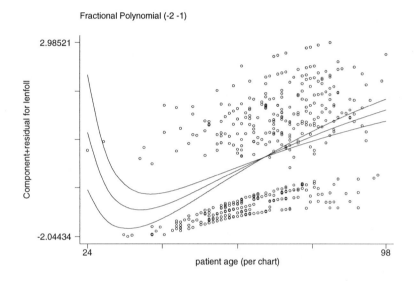

The fracplot command plots the partial residuals and the fitted fractional polynomial adjusting for other covariates, with 95% confidence limits, from the most recently fitted fractional polynomial model. The default is the best model, in this case the m = 2 model; i.e., a model with two power terms. This plot demonstrates how a few observations in either tail may effect the factional polynomial model. Here we see how two subjects younger than 30 (age=24 and 28) cause the drastic upturn in the plotted function. Aside from these two subjects the remainder of the plot is essentially linear. The choice is a linear model, which we show via the fracplot command below.

As shown in the graph, data fall around a straight line, supporting treating age as linear in the model.

. `fracpoly cox lenfol age, dead(fstat) compare degree(1)`
   `degree(#) determines the degree of fractional polynomial to be fitted.`

. fracplot

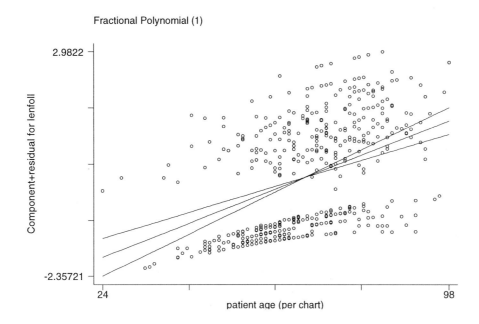

Fractional Polynomial (1)

Smoothed Residual Plot

Shown in the figure below is the smoothed residual based plot computed using the program listed below the figure. In the plot we see the same upward curve seen in the fractional polynomial. Rerunning the plot excluding the youngest subjects yields a nearly linear plot.

The STATA program to perform the smoothed residual plot is listed below. Use and modify as you wish. The program is not a distributed STATA command and as such does not have the options fully supported commands do, such as allowing missing data.

```
*!do file tp produce the GTF smoothed diagnostic plots
*! version of 6Sept01 by DWH
program define gtf_plot
version 7
tempvar  E y N_sm E_sm  M_sm res_sm ratio
tempname b
args N M x
gen `E' = `N' - `M'
scalar `b' =_b[`x']
ksm `N' `x', lowess gen(`N_sm') l1title("Observed Event")
more
ksm `E' `x', lowess gen(`E_sm') l1title("Expected Event")
more
quietly {gen `y' = ln(`N_sm'/`E_sm')+`b'*`x'}
sort `x'
graph `y'   `x' , l1title("ln(N_sm/E_sm)+beta*x") c(l) s(i) saving(`3'1,replace)
more
end
exit
```

use is as follows:
    gft_plot c M_t age
    `1' = censoring variable
    `2' = martingale residuals from cox model containing covariate
    `3' = covariate of interest

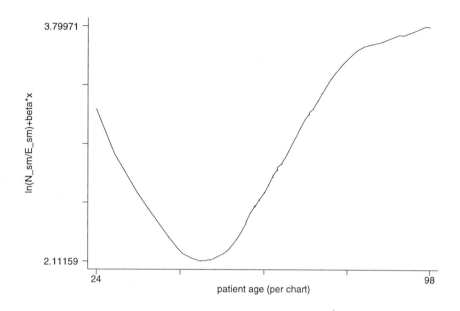

2. *Using the methods for model building discussed in this chapter, find the best model for estimating the effect of the covariates on long term survival following hospitalization for an acute myocardial infarction in the WHAS. This process should include the following steps: variable selection, assessment of the scale of continuous variables, and selection of interactions.*

Step 1: Variable selection

(i) Univariate Model

```
. stcox age, nohr

Cox regression -- entry time 0

No. of subjects =         481              Log likelihood = -1385.5758
No. of failures =         249              chi2(1)        =      70.07
Time at risk    =      834555              Prob > chi2    =     0.0000

   lenfol |
    fstal |    Coef.   Std. Err.      z     P>|z|    [95% Conf. Interval]
----------+------------------------------------------------------------
      age |  .0438179   .0053013    8.265   0.000    .0334275    .0542082
```

```
. stcox sex, nohr
Cox regression -- entry time 0

No. of subjects =         481           Log likelihood =  -1416.0443
No. of failures =         249           chi2(1)        =        9.13
Time at risk    =      834555           Prob > chi2    =      0.0025

  lenfol |
   fstal |      Coef.   Std. Err.       z     P>|z|    [95% Conf. Interval]
---------+----------------------------------------------------------------
     sex |   .3869737   .1271376     3.044    0.002    .1377886    .6361588
```

```
. stcox cpk, nohr

       failure _d:  fstat
   analysis time _t:  lenfol

Cox regression -- Breslow method for ties

No. of subjects =         481           Number of obs  =         481
No. of failures =         249
Time at risk    =      834555
                                        LR chi2(1)     =        0.31
Log likelihood  =  -1420.4525           Prob > chi2    =      0.5766

      _t |
      _d |      Coef.   Std. Err.       z     P>|z|    [95% Conf. Interval]
---------+----------------------------------------------------------------
     cpk |   .0000345   .0000609      0.57    0.571   -.0000848    .0001538
```

```
. stcox sho, nohr
Cox regression -- entry time 0

No. of subjects =         481           Log likelihood =  -1371.9345
No. of failures =         249           chi2(1)        =       97.35
Time at risk    =      834555           Prob > chi2    =      0.0000

  lenfol |
   fstal |      Coef.   Std. Err.       z     P>|z|    [95% Conf. Interval]
---------+----------------------------------------------------------------
     sho |   2.382395   .1959875    12.156    0.000    1.998266    2.766523
```

```
. stcox chf, nohr
     failure time:  lenfol
   failure/censor:  fstal

Cox regression -- entry time 0

No. of subjects =         481           Log likelihood =  -1387.5657
No. of failures =         249           chi2(1)        =       66.09
Time at risk    =      834555           Prob > chi2    =      0.0000

  lenfol |
   fstal |      Coef.   Std. Err.       z     P>|z|    [95% Conf. Interval]
---------+----------------------------------------------------------------
     chf |   1.047107   .1289397     8.121    0.000     .79439    1.299824
```

```
. stcox miord, nohr

    failure time:  lenfol
    failure/censor: fstat


Cox regression -- entry time 0

No. of subjects =         481            Log likelihood =  -1414.7302
No. of failures =         249            chi2(1)        =       11.76
Time at risk    =      834555            Prob > chi2    =      0.0006

   lenfol |
    fstal |    Coef.    Std. Err.       z     P>|z|    [95% Conf. Interval]
----------+------------------------------------------------------------------
    miord |  .4459343   .1282233     3.478    0.001     .1946212    .6972474
```

```
. tab mitype, gen(mi)

   mitype |    Freq.     Percent      Cum.
----------+-----------------------------------
        1 |      280       58.21      58.21
        2 |      195       40.54      98.75
        3 |        6        1.25     100.00
----------+-----------------------------------
    Total |      481      100.00
```

In the above table we see that there are only 6 subjects with mitype = 3, indeterminate. It is not wise to use a design variable based on so few subjects. We can either drop the subjects or pool them with one of the other two categories. We chose to pool them with mitype = 2 non-Qwave and use this dichotomous covariate. Other choices are possible and those with more clinical experience may feel another strategy is preferred.

```
. gen mitype01=mitype>1
```

```
. tab mitype mitype01

          |      mitype01
   mitype |     0          1  |   Total
----------+---------------------+----------
        1 |   280          0  |    280
        2 |     0        195  |    195
        3 |     0          6  |      6
----------+---------------------+----------
    Total |   280        201  |    481
```

```
. stcox mitype01, nolog nohr

failure _d:  fstat
   analysis time _t:  lenfol

Cox regression -- Breslow method for ties

No. of subjects =         481                Number of obs   =       481
No. of failures =         249
Time at risk    =      834555
                                             LR chi2(1)      =      1.50
Log likelihood  =   -1419.8579               Prob > chi2     =    0.2205

------------------------------------------------------------------------------
        _t |
        _d |      Coef.   Std. Err.      z    P>|z|     [95% Conf. Interval]
-------------+----------------------------------------------------------------
   mitype01 |  -.1586392   .1301794    -1.22   0.223    -.4137862    .0965077
------------------------------------------------------------------------------
```

All covariates except "cpk" are significant in the univariate analysis at the 25% level. Thus, we begin with a multivariable model that contains all covariates, but cpk.

(ii). Multivariate Model

```
. stcox age sex sho chf miord mitype01, nohr nolog noshow

Cox regression -- Breslow method for ties

No. of subjects =         481                Number of obs   =       481
No. of failures =         249
Time at risk    =      834555
                                             LR chi2(6)      =    175.23
Log likelihood  =   -1332.9922               Prob > chi2     =    0.0000

------------------------------------------------------------------------------
        _t |
        _d |      Coef.   Std. Err.      z    P>|z|     [95% Conf. Interval]
-------------+----------------------------------------------------------------
       age |   .0338543   .0058853     5.75   0.000     .0223194    .0453893
       sex |   .0385665   .1341826     0.29   0.774    -.2244264    .3015595
       sho |   1.802071   .2069908     8.71   0.000     1.396376    2.207765
       chf |    .568941    .142938     3.98   0.000     .2887876    .8490945
     miord |   .2937397   .1324702     2.22   0.027     .0341028    .5533765
  mitype01 |  -.2391096   .1339215    -1.79   0.074     -.501591    .0233718
------------------------------------------------------------------------------
```

Examining the p-values for the Wald statistics with the goal of trying to simplify the model, we see that the sex is not significant covariate. Thus, we fit a model excluding sex to see the change in the coefficients for the variables remaining in the model.

```
. stcox age sho chf miord mitype01, nohr nolog noshow

Cox regression -- Breslow method for ties

No. of subjects =          481                  Number of obs   =        481
No. of failures =          249
Time at risk    =       834555
                                                LR chi2(5)      =     175.15
Log likelihood  =   -1333.0335                  Prob > chi2     =     0.0000

------------------------------------------------------------------------------
      _t |
      _d |      Coef.   Std. Err.      z    P>|z|     [95% Conf. Interval]
---------+--------------------------------------------------------------------
     age |   .0342357   .0057344     5.97   0.000     .0229965    .0454748
     sho |   1.807153   .206332      8.76   0.000     1.402749    2.211556
     chf |   .5700533   .142831      3.99   0.000     .2901098    .8499969
   miord |   .2947876   .1323731     2.23   0.026     .0353411    .5542341
 mitype01|  -.2331995   .1323259    -1.76   0.078    -.4925536    .0261545
------------------------------------------------------------------------------
```

(iii) Is sex a confounder?

$$\Delta \hat{\beta}\% = 100 \frac{\hat{\theta} - \hat{\beta}}{\hat{\beta}},$$

where $\hat{\theta}$ denotes the crude estimator from the model that does not contain the potential confounder (sex) and $\hat{\beta}$ denotes the adjusted estimator from the model that does include the sex.

The maximum change in the coefficient for any variable remaining in the model is less than 15% thus we exclude sex from the model.

(iv). Put the "cpk" covariate back in the model

```
. stcox age sho chf miord mitype01 cpk, nohr nolog noshow

Cox regression -- Breslow method for ties

No. of subjects =          481                  Number of obs   =        481
No. of failures =          249
Time at risk    =       834555
                                                LR chi2(6)      =     176.00
Log likelihood  =   -1332.6085                  Prob > chi2     =     0.0000

------------------------------------------------------------------------------
      _t |
      _d |      Coef.   Std. Err.      z    P>|z|     [95% Conf. Interval]
---------+--------------------------------------------------------------------
     age |   .0349656   .0057842     6.05   0.000     .0236288    .0463024
     sho |   1.775766   .2090758     8.49   0.000     1.365985    2.185547
     chf |   .5752303   .1427333     4.03   0.000     .2954782    .8549825
   miord |   .3081254   .1331384     2.31   0.021     .0471789    .5690719
 mitype01|  -.2036522   .1362924    -1.49   0.135    -.4707804    .0634759
     cpk |   .0000631   .0000668     0.94   0.345    -.0000678    .000194
------------------------------------------------------------------------------
```

66 CHAPTER 5 SOLUTIONS

We see that cpk is not significant in the multivariate model. Thus, the preliminary main effect model has five variables: age, sho, chf, miord, and mitype01.

The next step for modeling is involved in an assessment of the scale of continuous variables. Among the variables in the preliminary main effect model, age is a continuous covariate. The linearity assumption needs to be checked for the age covariate.

Step 2: Assessment of the scale of continuous variables

(i). To check the scale in the multivariable model we use fractional polynomials and the smoothed residual plot. The results of the fractional polynomial analysis are shown next.

```
. fracpoly cox lenfol age sho chf miord mitype01, nohr nolog dead(fstat) compare

Model fit output is not shown.

Fractional polynomial model comparisons:
-----------------------------------------------------------------
age              df       Deviance      Gain    P(term)  Powers
-----------------------------------------------------------------
Not in model     0        2702.187      --       --
Linear           1        2666.067      0.000    0.000    1
m = 1            2        2665.603      0.464    0.496    2
m = 2            4        2660.694      5.373    0.086   -1 -1
-----------------------------------------------------------------
```

. **fracplot**

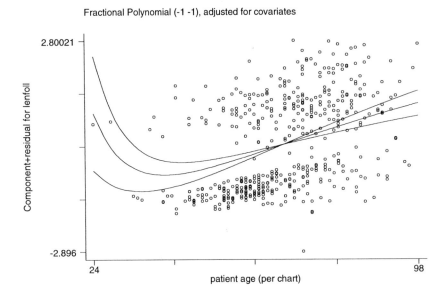

The results show that, except for the effect of the two young subjects discussed in problem 2, the model is linear in age.

The plot shown below again shows the effect of the few young subjects on the shape of the smoothed plot.

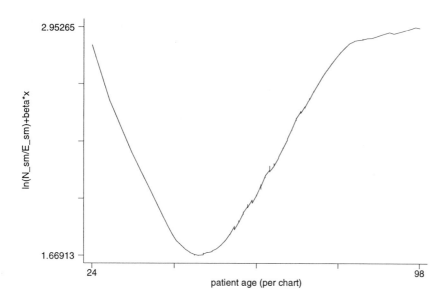

Analysis with the youngest two subjects deleted yielded non-significant fractional polynomials and a much more modest upturn in the smoothed residual plot. We decide to model age as linear in the log hazard.

Step 3: Selection of interactions

The steps in selection interactions are as follows:
    1. Consult with the study team to determine the pair of variables clinically plausible for interaction.
    2. Interaction terms should be statistically significant.
    3. Added interactions to the main effects model and simplify using Wald statistics.

Note that all main effect terms must remain in the model regardless of their significance. What is important is how the main effect and the significant interaction terms combine to estimate hazard ratios of interest.

For sake of demonstration we considered all possible interactions. Among these, three were significant at the 5 percent level. The results of generating and testing these three are shown below.

CHAPTER 5 SOLUTIONS

```
. gen shoxchf = sho*chf
```

```
. stcox   age sho chf miord mitype01 shoxchf, nohr nolog noshow

Cox regression -- Breslow method for ties

No. of subjects =         481                Number of obs   =       481
No. of failures =         249
Time at risk    =      834555
                                             LR chi2(6)      =    179.92
Log likelihood  =  -1330.6486                Prob > chi2     =    0.0000

------------------------------------------------------------------------------
         _t |
         _d |     Coef.   Std. Err.      z    P>|z|    [95% Conf. Interval]
------------+-----------------------------------------------------------------
        age |   .0329628   .0057617    5.72   0.000      .02167    .0442555
        sho |   3.128423   .5349019    5.85   0.000    2.080034    4.176811
        chf |   .6335305   .1448123    4.37   0.000    .3497036    .9173574
      miord |    .322305   .1331705    2.42   0.016    .0612956    .5833144
   mitype01 |  -.2123587   .1330683   -1.60   0.111   -.4731678    .0484504
    shoxchf |  -1.430156   .5645232   -2.53   0.011   -2.536601   -.3237104
------------------------------------------------------------------------------

. lrtest,saving(0)
```

```
. quietly stcox age sho chf miord mitype01, nohr nolog noshow
```

```
. lrtest
Cox:   likelihood-ratio test                    chi2(1)    =      4.77
                                                Prob > chi2 =    0.0290
```

```
. gen shoxmiord=sho*miord

. stcox   age sho chf miord mitype01 shoxmiord, nohr nolog noshow

Cox regression -- Breslow method for ties

No. of subjects =         481                Number of obs   =       481
No. of failures =         249
Time at risk    =      834555
                                             LR chi2(6)      =    184.61
Log likelihood  =  -1328.3021                Prob > chi2     =    0.0000

------------------------------------------------------------------------------
         _t |
         _d |     Coef.   Std. Err.      z    P>|z|    [95% Conf. Interval]
------------+-----------------------------------------------------------------
        age |   .0328211    .005773    5.69   0.000    .0215061    .0441361
        sho |   2.427992   .2677602    9.07   0.000    1.903192    2.952793
        chf |   .5895391   .1422212    4.15   0.000    .3107907    .8682875
      miord |   .4574347   .1402849    3.26   0.001    .1824814    .7323881
   mitype01 |  -.2172837   .1324024   -1.64   0.101   -.4767877    .0422203
  shoxmiord |  -1.120981   .3653705   -3.07   0.002   -1.837094   -.4048682
------------------------------------------------------------------------------

. lrtest,saving(0)
```

```
. quietly stcox age sho chf miord mitype01, nohr nolog noshow
```

```
. lrtest
Cox:  likelihood-ratio test                       chi2(1)    =        9.46
                                                  Prob > chi2 =      0.0021
```

```
. gen shoxmityp=sho*mitype01
```

```
. stcox   age sho   chf miord mitype01 shoxmityp, nohr nolog noshow

Cox regression -- Breslow method for ties

No. of subjects =          481                    Number of obs   =        481
No. of failures =          249
Time at risk    =       834555
                                                  LR chi2(6)      =     179.40
Log likelihood  =   -1330.9103                    Prob > chi2     =     0.0000

------------------------------------------------------------------------------
       _t |
       _d |      Coef.   Std. Err.       z     P>|z|     [95% Conf. Interval]
----------+-------------------------------------------------------------------
      age |   .0344763   .0057439      6.00    0.000     .0232184    .0457341
      sho |   2.090218   .2389522      8.75    0.000      1.62188    2.558556
      chf |   .5785631   .1426689      4.06    0.000     .2989373     .858189
    miord |    .302203   .1318092      2.29    0.022     .0438618    .5605443
 mitype01 |  -.1240558   .1416571     -0.88    0.381    -.4016985     .153587
shoxmityp |  -.7893241   .3977963     -1.98    0.047    -1.568991   -.0096576
------------------------------------------------------------------------------

. lrtest,saving(0)
```

```
. quietly stcox age sho chf miord mitype01, nohr nolog noshow
```

```
. lrtest
Cox:  likelihood-ratio test                       chi2(1)    =        4.25
                                                  Prob > chi2 =      0.0393
```

One potential problem is that there are only 38 subjects with shock (sho = 1). As a result the significant interactions could be due to few subjects. Put in all significant interaction terms found previously

```
. stcox  age sho   chf miord mitype01 shoxchf shoxmiord shoxmityp, nohr
nolog noshow

Cox regression -- Breslow method for ties

No. of subjects =         481                Number of obs   =       481
No. of failures =         249
Time at risk    =      834555
                                             LR chi2(8)      =    188.41
Log likelihood  =  -1326.4035                Prob > chi2     =    0.0000

------------------------------------------------------------------------------
        _t |
        _d |     Coef.   Std. Err.      z    P>|z|     [95% Conf. Interval]
-------------+----------------------------------------------------------------
       age |   .0322924      .0058     5.57   0.000     .0209247    .0436602
       sho |   3.254533    .5382089    6.05   0.000     2.199663    4.309403
       chf |   .6379368    .1451732    4.39   0.000     .3534026    .922471
     miord |   .4450963    .1405126    3.17   0.002     .1696968    .7204959
   mitype01|  -.1392654    .1418638   -0.98   0.326    -.4173132    .1387825
    shoxchf|  -.8216494    .5851016   -1.40   0.160    -1.968428    .3251287
  shoxmiord|  -.9084194    .3863107   -2.35   0.019    -1.665574   -.1512644
  shoxmityp|  -.5127787    .4130222   -1.24   0.214    -1.322287     .29673
------------------------------------------------------------------------------

. lrtest, saving(0)
```

We note that the interactions of chf and mitype01 with sho are no longer significant so we exclude them from the model and refit.

```
. stcox  age sho   chf miord mitype01 shoxmiord , nohr nolog noshow

Cox regression - Breslow method for ties

No. of subjects =         481                Number of obs   =       481
No. of failures =         249
Time at risk    =      834555
                                             LR chi2(6)      =    184.61
Log likelihood  =  -1328.3021                Prob > chi2     =    0.0000

------------------------------------------------------------------------------
        _t |
        _d |     Coef.   Std. Err.      z    P>|z|     [95% Conf. Interval]
-------------+----------------------------------------------------------------
       age |   .0328211    .005773     5.69   0.000     .0215061    .0441361
       sho |   2.427992   .2677602     9.07   0.000     1.903192    2.952793
       chf |   .5895391   .1422212     4.15   0.000     .3107907    .8682875
     miord |   .4574347   .1402849     3.26   0.001     .1824814    .7323881
   mitype01|  -.2172837   .1324024    -1.64   0.101    -.4767877    .0422203
  shoxmiord|  -1.120981   .3653705    -3.07   0.002    -1.837094   -.4048682
------------------------------------------------------------------------------

. lrtest
Cox:  likelihood-ratio test                        chi2(2)    =     3.80
                                                   Prob > chi2 =   0.1498
```

The likelihood ratio test is not significant. We note that the *p*-value for the Wald statistic for the sho by miord interaction is less than 0.05 so we retain this single interaction in the model

The results for the smaller model containing a single interaction indicate that the Wald statistics for dichotomized covariate mitype01 is no longer significant at the 10 percent level, $p = 0.101$. However, evaluating the partial likelihood ratio test we find $p = 0.098$. The variable mitype01 is not a strong confounder; but since it is marginally significant we decide to keep it. One could argue that it should be deleted since $p > 0.05$ and it is not a confounder. In practice the physicians on the study team would have to give a ruling as to its subject matter relevance.

Thus we select as our preliminary final model the one shown below containing, age chf, sho, miord, mitype01 and shoxmiord.

```
. stcox age chf sho miord mitype01 shoxmiord ,nohr noshow nolog

Cox regression -- Breslow method for ties

No. of subjects =         481                Number of obs   =        481
No. of failures =         249
Time at risk    =      834555
                                             LR chi2(6)      =     184.61
Log likelihood  =   -1328.3021               Prob > chi2     =     0.0000

------------------------------------------------------------------------------
         _t |
         _d |     Coef.    Std. Err.      z     P>|z|    [95% Conf. Interval]
------------+-----------------------------------------------------------------
        age |  .0328211    .005773      5.69   0.000     .0215061    .0441361
        sho |  2.427992    .2677602     9.07   0.000     1.903192    2.952793
        chf |  .5895391    .1422212     4.15   0.000     .3107907    .8682875
      miord |  .4574347    .1402849     3.26   0.001     .1824814    .7323881
   mitype01 | -.2172837    .1324024    -1.64   0.101    -.4767877    .0422203
  shoxmiord | -1.120981    .3653705    -3.07   0.002    -1.837094   -.4048682
------------------------------------------------------------------------------
```

3. *Present the results of the model selected in problem 2 in a table or tables that are suitable for publication in an applied journal. This presentation should include estimates of hazard ratios, with confidence intervals.*

We present results here as if we had evaluated the fit of the model, influence of individual subjects and adherence to the proportional hazards assumption. These topics are considered in Chapter 6 and we evaluate the model in problem 5.

Table 4.1. Estimated Hazard Ratios and 95 Percent Confidence Intervals from the Preliminary Final Model from the WHAS.

Variable	Hazard Ratio	95 % CI
Age*	1.39	1.24, 1.55
Congestive Heart Failure	1.80	1.36, 2.38
Non Q-wave	0.80	0.62, 1.04
Shock		
No: Recurrent MI	1.58	1.20, 2.08
Yes: Recurrent MI	0.52	0.27, 1.00
MI Order		
First: Shock	11.34	6.71, 19.16
Recurrent: Shock	3.69	2.11, 6.46

*: Hazard ratio for a 10 year increase in age

The estimated hazard ratios for MI order within level of shock and for shock within level of MI order were obtained using STATA's lincom feature. For example, e.g, for shock within MI recurrent,

```
. lincom _b[sho]+_b[shoxmiord],hr

 ( 1)  sho + shoxmiord = 0.0

------------------------------------------------------------------------------
      _t | Haz. Ratio   Std. Err.      z    P>|z|     [95% Conf. Interval]
---------+--------------------------------------------------------------------
     (1) |   3.695113   1.053172     4.59   0.000     2.113588    6.460038
------------------------------------------------------------------------------

. lincom _b[miord]+_b[shoxmiord],hr

 ( 1)  miord + shoxmiord = 0.0

------------------------------------------------------------------------------
      _t | Haz. Ratio   Std. Err.      z    P>|z|     [95% Conf. Interval]
---------+--------------------------------------------------------------------
     (1) |   .5150215   .1744118    -1.96   0.050     .2651957    1.000194
------------------------------------------------------------------------------
```

Chapter Six – Solutions

1. Using data from the HMO-HIV+ study, assess the fit the proportional hazards model containing AGE and DRUG. This assessment of fit should include the following steps: evaluation of the proportional hazards assumption for each of the two covariates, examination of diagnostic statistics, and an overall test of fit. If the model does not fit or adhere to the proportional hazards assumption what would you do next? Note: the goal is to obtain a model to estimate the effect of AGE and DRUG on the survivorship experience.

The lack of an obvious residual has lead to the development of several different residuals. Among them, Stata offers score residuals (**esr**), martingale residuals (**mgale**), Schoenfeld residuals (**schoenfeld**), and scaled Schoenfeld residuals (**scaledsch**) with options specified to the **stcox** command. Cox-Snell and deviance residuals can be calculated with options to **predict**. Although the uses of residuals vary and depend on user preferences, Schoenfeld and scaled Schoenfeld residuals are frequently used for testing the proportional hazard assumption, while score residuals are used for examining leverage points and identifying outliers.

We begin by fitting the model and saving the martingale, Schoenfeld, scaled Schoenfeld and score residuals for later use. Following the fit of the model we calculate, using matrix commands, the scaled score residuals that yield influence measures for the two estimated coefficients.

```
. stcox  AGE DRUG, nolog nohr  schoenfeld(sc*)  scaledsch(sca*)   esr(scr*)
mgale(M_t)

       failure _d:  CENSOR
   analysis time _t:  TIME

Cox regression -- Breslow method for ties

No. of subjects =          100                Number of obs   =        100
No. of failures =           80
Time at risk    =         1136
                                              LR chi2(2)      =      34.98
Log likelihood  =   -281.70404                Prob > chi2     =     0.0000

------------------------------------------------------------------------------
    _t |
    _d |      Coef.   Std. Err.      z    P>|z|     [95% Conf. Interval]
-------+----------------------------------------------------------------------
   AGE |   .0915319   .0184879     4.95   0.000     .0552963    .1277675
  DRUG |   .9413856   .2555104     3.68   0.000     .4405943    1.442177
------------------------------------------------------------------------------
```

The following STATA do file is used to calculate the scaled score residuals and likelihood displacement statistic or Cook's distance for proportional hazards models. The reader may modify it as needed.

```
* Do file sscr.do
* program to compute the scaled score residuals in eq(6.22) page 219
* and Cooks distance, ld in eq (6.24) page 221
capture program drop sscr
program define sscr
args p
capture drop ld
local j = 1
while `j'<= `p' {
capture drop sscr`j'
local j = `j'+1
}
mkmat scr1-scr`p', matrix(scr)
mat V=e(V)
mat inf=scr*V
svmat inf, names(sscr)
local j = 1
while `j'<= `p' {
label var sscr`j' "dfbeta_x`j'"
local j = `j'+1
}
mat LDbig=inf*scr'
mat LD=vecdiag(LDbig)'
svmat LD, names(ld)
rename ld1 ld
label var ld "Cook_dist"
mat drop LDbig
mat drop LD
mat drop inf
mat drop scr
end
sscr `*'
exit
* use do sscr 10
    `1' = number of covariates in the model
```

(1) Evaluation of the proportional hazards assumption for the covariates.

(1.a) Numerical methods using scaled Schoenfeld residuals

To perform global test (**stphtest**), the **schoenfeld(sc*)** option should be specified when **stcox** is estimated. Schoenfeld residuals are based on the individual contributions to the derivative of the log-partial likelihood. Since we have two covariates, two Schoenfeld residuals will be created. Stata sets the value of the estimate of the Schoenfeld residual to missing for subjects whose observed survival time is censored. The **scaledsch(ssc*)** option is to calculate the scaled Schoenfeld residuals where **mgale** option is to calculate the martingale residuals.

The **stphtest** is a test of nonzero slope in a generalized linear regression of the scaled Schoenfeld residuals on functions of time. The rejection of the null hypothesis of a zero slope indicates that the proportional hazards assumption is violated.

```
. stphtest, log detail            log indicates log(time)

  Test of proportional hazards assumption
      Time:   Log(t)
  ----------------------------------------------------------------
                  |      rho         chi2        df      Prob>chi2
  ----------------+-----------------------------------------------
       AGE        |    0.05068       0.18         1        0.6726
       DRUG       |    0.01264       0.01         1        0.9118
  ----------------+-----------------------------------------------
       global test|                  0.18         2        0.9141
  ----------------------------------------------------------------
```

The **stphtest** supports that both covariates have zero slopes. No evidence to violated the proportional hazards assumption

We note that in STATA version it is possible to explicitly test for zero coefficient in the time-varying coefficient model. The results and STATA commands are shown next.

```
. stcox AGE DRUG, nolog nohr noshow tvc(DRUG AGE) texp(ln(_t))

Cox regression -- Breslow method for ties

No. of subjects =          100                  Number of obs   =        100
No. of failures =           80
Time at risk    =         1136
                                                LR chi2(4)      =      35.16
Log likelihood  =   -281.61464                  Prob > chi2     =     0.0000

------------------------------------------------------------------------------
          _t |
          _d |     Coef.   Std. Err.      z    P>|z|    [95% Conf. Interval]
-------------+----------------------------------------------------------------
rh           |
         AGE |  .0811645   .0309948     2.62   0.009     .0204159    .1419132
        DRUG |  .8916303   .4560034     1.96   0.051     -.00212     1.78538
-------------+----------------------------------------------------------------
t            |
        DRUG |  .0401428   .267503      0.15   0.881    -.4841535    .5644391
         AGE |  .0065743   .0156238     0.42   0.674    -.0240477    .0371964
------------------------------------------------------------------------------
note: second equation contains variables that continuously vary with respect to
time; variables are interacted with current values of ln(_t).
```

Note that the p-values for the interaction of DRUG and AGE with ln(_t) are not significant, further supporting the assumption of proportional hazards.

(1.b) Graphical methods using the scaled Schoenfeld residual plots. We obtain these from the sphtest plot option.

AGE

`. stphtest, log plot(AGE)`

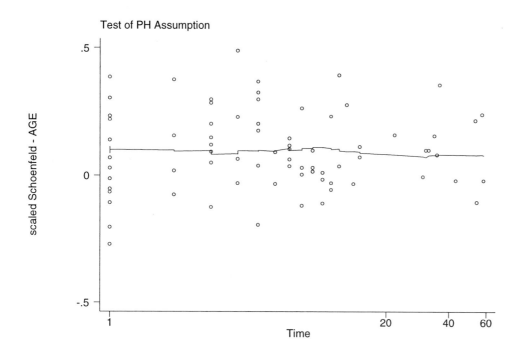

No systematic departure of residual as a function of time. The polygon connecting the values of the smoothed residuals has approximately a zero slope.

DRUG

```
.stphtest , log plot(DRUG)
```

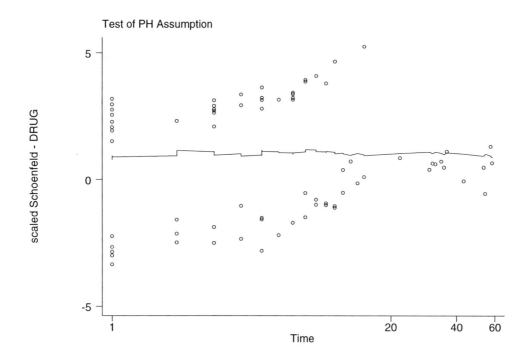

No systematic departure of residual as a function of time. The polygon connecting the values of the smoothed residuals has approximately a zero slope.

(2) Examination of diagnostic statistics

(2.a) Score residual

The score residuals are sometimes referred to as the "leverage" or "partial leverage" residuals. The graphs of the score residuals for the covariates of AGE and Drug are as follows.

AGE

Since age is a continuous variable, it best to plot the score residuals versus the values of age. The score residuals for age display the fan shape. The band is narrowest at the mean age of 36 years and increases for ages getting older or younger than 36. There are two possible high leverage points, but not too distant from the other values.

```
. graph   scr1 AGE
```

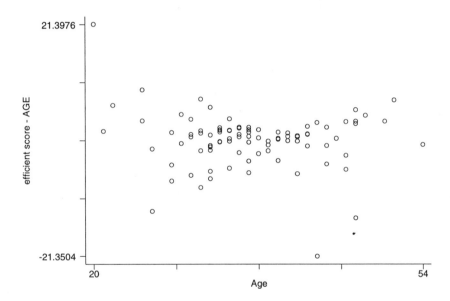

Drug

The best way to display the score residuals for a dichotomous or polychotomous covariate is by a box and whisker plot. The plots indicate two large negative values for DRUG = 1.

```
. graph  scr2,  box  by(DRUG )
```

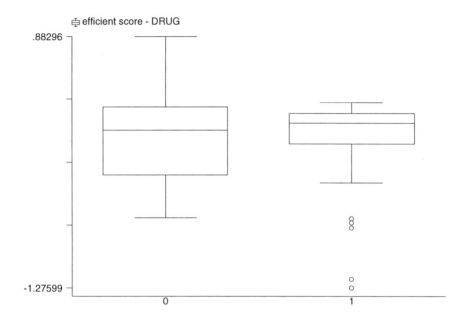

(2.b) Scaled score residual (dbeta)
The influence plots are the scaled score residuals calculated earlier using the matrix commands in STATA.

AGE
There are no distinct breaks in the points in fan shape. Although there is one subject that lies a bit below the fan with age about 42 and one whose age is 20.

`. graph sscr1 AGE`

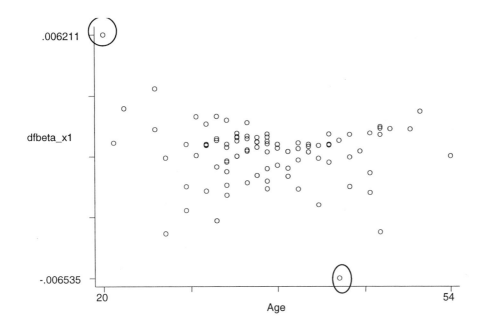

Drug

The box plot looks quite similar to the previous one and identifies two subjects that are potentially influential with DRUG = 1.

`. graph  sscr2, box by(DRUG )`

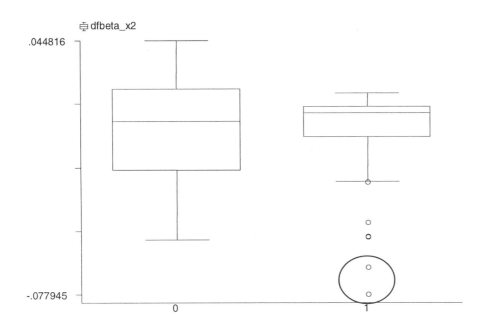

Next we plot the overall influence diagnostic, *ld*, versus the martingale residuals

`. graph ld mgale`

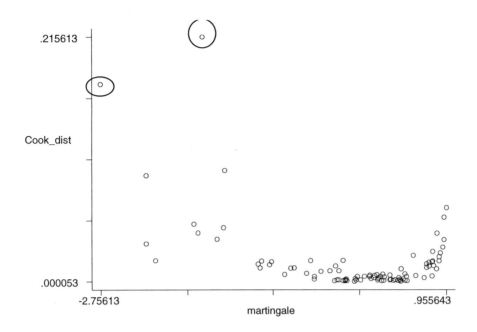

The plot shows two subjects, top left corner, that lie well away from the rest of the points.

When we refit the model deleting these two subjects the percent change in both coefficients is less than 10 percent. Thus we conclude that they are not overly influential in determining the estimates.

(3) Overall test of fit

We use the partial likelihood ratio test to perform the Grønnesby-Borgan test. This procedure has been implemented into a small STATA do file that performs the test and calculates statistics used to form a table of observed and expected numbers of events within each risk group. Finally the program obtains the Arjas plots within each risk group. We list the do file and then show its results for the current model.

In the HMO-HIV Study there are 80 deaths, non-censored observations. If we use 10 risk groups then we would expect about 8 deaths in each group. This is a bit too small to rely on the normal approximation to the Poisson distribution thus we use 8 groups to obtain expected values of a bout 10.

```
*! Do file survfit.do
* do file to do Arjas plots
* and the Gronnesby-Borgan test
* File will handle stcox and all streg models
* Version of 9/12/01 DWH
args  M_t G
capture drop H_t
capture drop Exp
capture drop Obs
capture drop r
capture drop group
capture drop zgroup
capture drop pzgroup
mat btht=get(_b)
mat bcoef=btht[1,1..e(df_m)]
global  S_xvar1 : colnames bcoef
mat drop btht
mat drop bcoef
gen H_t= _d -`M_t'
predict r,xb
sort r
gen group=group(`G')
sort group r
by group: gen Exp=sum(H_t)
by group: gen Obs=sum(_d)
by group: gen zgroup=cond(_n==_N, (Obs-Exp)/sqrt(Exp),.)
by group: gen pzgroup=cond(_n==_N,2*(1-norm(abs(zgroup))),.)
graph Exp Obs Obs if(_d==1), c(ll) s(ii) by(group) /*
    */ saving(c:\empty\Arj_$S_E_cmd,replace) b1title("Arjas plot for $S_E_cmd model")

more
quietly xi:$S_E_cmd _t $S_xvar1 i.group, dead(_d)
lrtest , saving(0)
quietly xi:$S_E_cmd _t $S_xvar1, dead(_d)
lrtest
sort group
by group : list Obs Exp zgroup pzgroup if _n==_N
exit

* use do survfit M_t 10

      M_t = martingale residuals
      10 = number of groups
```

CHAPTER 6 SOLUTIONS

```
. do survfit M_t 8

Cox:   likelihood-ratio test                    chi2(7)     =      2.97
                                                Prob > chi2 =    0.8880
. sort group

. by group : list Obs Exp zgroup pzgroup if _n==_N


-> group = 1

             Obs         Exp        zgroup      pzgroup
  13.          8     10.57894    -.7929041     .4278337

-> group = 2

             Obs         Exp        zgroup      pzgroup
  25.         11     8.776608     .7505036     .4529515

-> group = 3

             Obs         Exp        zgroup      pzgroup
  38.         11      11.6916    -.2022645     .8397099

-> group = 4

             Obs         Exp        zgroup      pzgroup
  50.          9     6.820933     .8343509     .4040833

-> group = 5

             Obs         Exp        zgroup      pzgroup
  62.         10     9.482466     .1680655     .8665318

-> group = 6

             Obs         Exp        zgroup      pzgroup
  75.         10     10.32922    -.1024347     .9184116

-> group = 7

             Obs         Exp        zgroup      pzgroup
  87.         12     11.03493     .2905183     .7714198

-> group = 8

             Obs         Exp        zgroup      pzgroup
 100.          9      11.2853    -.6802787     .496328
.exit

end of do-file
```

The results of the test are not significant, p= 0.880, thus we can not reject the null hypothesis that the model fits. The next step is to form the 2 by 8 table of observed and expected events and the within decile z-score, similar to Table 6.5 page 228. We leave re-expressing the results from the STATA print out to the reader. The results from the STATA do file support model fit within each octile of risk.

We note that are ties in the risk score. The above procedure just breaks the data into 8 equal sized groups, possible splitting subjects with the same risk score into different groups. The value of the test will depend somewhat on how events and censored observations are split up. If the model fits then the test should not be significant; but the actual values may be slightly different depending on the grouping.

Next we show the Arjas plots by octile of risk score obtained from survfit.do.

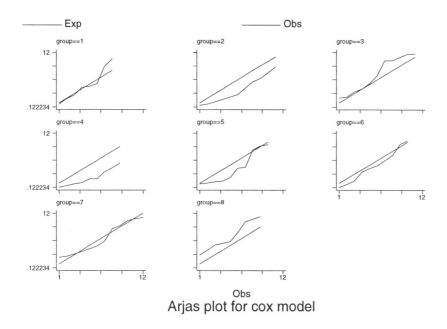

Arjas plot for cox model

The plot of expected versus observed follows the 45 degree line in some but not all octiles of risk. We note that the z statistics and their associated p-values support fit. However these tests are based on the last point plotted in each figure. Some of the deviation from the 45-degree line could be due to small numbers of subjects in each group. We leave it to the reader to rerun survfit.do using 4 groups.

86 CHAPTER 6 SOLUTIONS

2. *Using the model obtained at the conclusion of problem 1, present a table of estimated hazard ratios, with confidence intervals. Present graphs of the age-adjusted, at the mean age, estimated survivorship functions for the two drug use groups. Use the estimated survivorship functions to estimate the age-adjusted median survival time for each of the two drug use groups.*

(a). Estimated hazard ratios with confidence intervals

Variable	HR	95 % CI
AGE	2.50	1.73, 3.59
DRUG	2.56	1.55, 4.22

*: Hazard Ratio for a 10 year increase in age

(b). Present graphs of the age-adjusted, at the mean age, estimated survivorship functions for the two drug use groups.

```
. quietly sum age
. gen AGE_c=AGE-r(mean)

. stcox AGE_c  DRUG, nolog nohr bases(S_base)

        failure _d:  CENSOR
   analysis time _t:  TIME

Cox regression -- Breslow method for ties

No. of subjects =           100              Number of obs   =        100
No. of failures =            80
Time at risk    =          1136
                                             LR chi2(2)      =      34.98
Log likelihood  =    -281.70404              Prob > chi2     =     0.0000

------------------------------------------------------------------------------
        _t |
        _d |      Coef.   Std. Err.      z    P>|z|     [95% Conf. Interval]
-----------+------------------------------------------------------------------
     AGE_c |   .0915319   .0184879     4.95   0.000     .0552963    .1277675
      DRUG |   .9413856   .2555104     3.68   0.000     .4405943    1.442177
------------------------------------------------------------------------------

. gen S0=S_base

. gen S1=S_base^(exp(_b[DRUG]))

. sort TIME

. graph S0 S1 TIME, yscale(0,1) s(ii) c(J[-]J[l]))
```

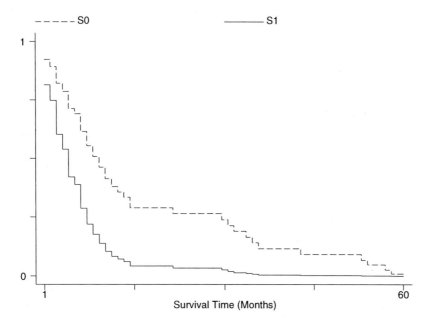

Survival Time (Months)

(c). Use the estimated survivorship functions to estimate the age-adjusted median survival time for each of the two drug use groups

```
. list TIME S0 S1 if 0.46< S0 & S0 < 0.9 & CENSOR==1

           TIME         S0          S1
 18.          2    .8930166    .7482159

Deleted lines from list

 40.          4    .7854212    .5383842
 42.          4    .7854212    .5383842
 43.          4    .7854212    .5383842
 44.          4    .7854212    .5383842
 45.          5    .7132945    .4205816    Median for DRUG = 1
 46.          5    .7132945    .4205816
 47.          5    .7132945    .4205816
 48.          5    .7132945    .4205816

Deleted lines from list

 66.          9    .5075026    .1757445
 67.          9    .5075026    .1757445
 68.          9    .5075026    .1757445
 69.         10    .4613784    .1376577    Median for DRUG = 0
 70.         10    .4613784    .1376577
 71.         10    .4613784    .1376577
```

Based on the above listing and applying the definition, $\hat{t}_{50} = \min(t, \hat{S}(t) \leq 0.5)$, we obtain estimates of 5 months for subjects with as history and 10 months for those without a history of IV drug use.

3. *In Section 6.4 diagnostic statistics were plotted and a few subjects were identified as being possibly influential. Fit the model shown in Table 6.6 deleting these subjects one at a time and then, collectively, calculate the percent change in all coefficients with each deletion. Do you agree or disagree with the conclusion in Section 6.4 to keep all subjects in the analysis? Explain the rationale for your decision.*

```
. stcox   age  becktota  ndrugfp1 ndrugfp2 ivhx_3 race treat site   agesite racesite ,
nohr nolog noshow esr(scr*) mgale(M_t)    This is the model presented in 6.6 and we
request that the score residuals be saved

Cox regression -- Breslow method for ties

No. of subjects =          575                    Number of obs    =        575
No. of failures =          464
Time at risk    =       138900
                                                  LR chi2(10)      =      67.13
Log likelihood  =    -2630.4179                   Prob > chi2      =     0.0000


------------------------------------------------------------------------------
         _t |
         _d |      Coef.   Std. Err.      z    P>|z|     [95% Conf. Interval]
------------+-----------------------------------------------------------------
        age |  -.0413963   .0099128    -4.18   0.000    -.060825   -.0219676
   becktota |   .008738    .0049654     1.76   0.078    -.000994    .0184701
   ndrugfp1 |  -.5744565   .1251879    -4.59   0.000    -.8198202  -.3290927
   ndrugfp2 |  -.2145782   .048587     -4.42   0.000    -.309807   -.1193494
     ivhx_3 |   .227748    .108563      2.10   0.036     .0149684   .4405276
       race |  -.4668853   .1347564    -3.46   0.001    -.7310029  -.2027677
      treat |  -.2467592   .09434      -2.62   0.009    -.4316621  -.0618562
       site | -1.316987    .5314407    -2.48   0.013   -2.358592   -.2753827
    agexsite|   .0324002   .0160807     2.01   0.044     .0008827   .0639178
    racexsit|   .8502773   .2477582     3.43   0.001     .3646801  1.335875
------------------------------------------------------------------------------
```

Scaled score residuals are used to identify influential observations.

The scaled score residuals are created using the previous STATA do file as follows:

(53 missing values generated)
. drop if xb==.
(53 observations deleted)

```
. sum xb

    Variable |       Obs        Mean    Std. Dev.       Min        Max
-------------+--------------------------------------------------------
          xb |       575   -2.214508    .4012434   -3.562604  -1.046997
```

```
. do sscr 10

. * Do file sscr.do
. * program to compute the scaled score residuals in
eq(6.22) page 219
. * and Cooks distance, ld in eq (6.24) page 221
. capture program drop sscr

. program define sscr
  1. args p
  2. capture drop ld
  3. local j = 1
  4. while `j'<= `p' {
  5. capture drop sscr`j'
  6. local j = `j'+1
  7. }
  8. mkmat scr1-scr`p', matrix(scr)
  9. mat V=e(V)
 10. mat inf=scr*V
 11. svmat inf, names(sscr)
 12. local j = 1
 13. while `j'<= `p' {
 14. label var sscr`j' "dfbeta_x`j'"
 15. local j = `j'+1
 16. }
 17. mat LDbig=inf*scr'
 18. mat LD=vecdiag(LDbig)'
 19. svmat LD, names(ld)
 20. rename ld1 ld
 21. label var ld "Cook_dist"
 22. mat drop LDbig
 23. mat drop LD
 24. mat drop inf
 25. mat drop scr
 26. end

. sscr `*'

. exit

end of do-file
```

AGE

. **graph sscr1 age**   *graph scaled score residuals for age to find largest one*

We have written a small STATA do file, listed below, to aid in the computation of the percent

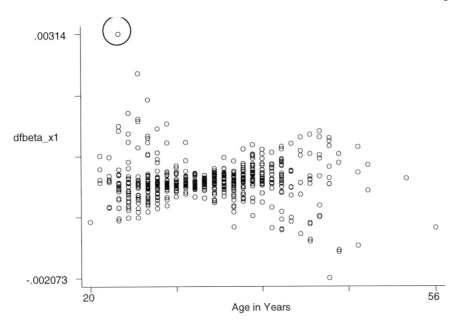

change in the coefficients

```
* File delta_beta.do
*Program written to compute delta beta % when models are fit deleting cases but keeping
* the variables the same
* must set bfull first using
*matrix bfull = get(_b)
*matrix V = get(VCE)
*matrix VI=syminv(V)

tempname balt bdif bfd bfdi pctchng cooksd
matrix `balt' = get(_b)
matrix `bdif' = `balt'-bfull
matrix `cooksd' = (`bdif')*(VI)*(`bdif'')
dis " Actual Cook's Distance"
mat list `cooksd'
matrix `bfd' = diag(bfull)
matrix `bfdi' = inv(`bfd')
mat `pctchng'=`bdif'*`bfdi'
mat `pctchng'=100*`pctchng'
dis "Percent Change in Coefficients from Full Data Model"
mat list `pctchng'
```

We begin the process by fitting the model excluding the point circled in the above graph. In this case we show all the output, but for other fits we only present the output containing the percent change in the coefficients.

```
. quietly stcox  age becktota ndrugfp1 ndrugfp2 ivhx_3 race treat site
agexsite racexsit, nolog nohr noshow

. mat bful=get(_b)

. mat bfull=get(_b)

. mat V=get(VCE)

. mat VI=syminv(V)

. stcox  age becktota ndrugfp1 ndrugfp2 ivhx_3 race treat site agexsite
racexsit if sscr1<0.003,nolog nohr noshow

Cox regression -- Breslow method for ties

No. of subjects =          574                 Number of obs   =       574
No. of failures =          464
Time at risk    =       138303
                                                LR chi2(10)     =     70.70
Log likelihood  =   -2626.9701                  Prob > chi2     =    0.0000

------------------------------------------------------------------------------
          _t |
          _d |      Coef.   Std. Err.      z    P>|z|     [95% Conf. Interval]
-------------+----------------------------------------------------------------
         age |  -.0447618   .0100707    -4.44   0.000    -.0644999   -.0250236
    becktota |   .0090366   .0049407     1.83   0.067    -.000647    .0187203
    ndrugfp1 |  -.5612852   .125035     -4.49   0.000    -.8063494   -.316221
    ndrugfp2 |  -.209473    .0485362    -4.32   0.000    -.3046022   -.1143438
      ivhx_3 |   .2509057   .1087286     2.31   0.021     .0378016    .4640098
        race |  -.4778407   .1346737    -3.55   0.000    -.7417964   -.213885
       treat |  -.2645226   .0945333    -2.80   0.005    -.4498045   -.0792407
        site |  -1.420371   .5334415    -2.66   0.008    -2.465897   -.3748447
    agexsite |   .0353458   .0161519     2.19   0.029     .0036887    .0670029
    racexsit |   .865281    .2478107     3.49   0.000     .3795809    1.350981
------------------------------------------------------------------------------
```

```
. do delta_beta

.* File delta_beta.do
.*Program written to compute delta beta % when models are fit deleting cases
*        but keeping the variables the same
.* must set bfull first using
. *matrix bfull = get(_b)
. *matrix V = get(VCE)
. *matrix VI=syminv(V)
.
. tempname balt bdif bfd bfdi pctchng cooksd
. matrix `balt' = get(_b)
. matrix `bdif' = `balt'-bfull
. matrix `cooksd' = (`bdif')*(VI)*(`bdif'')
. dis " Actual Cook's Distance"
 Actual Cook's Distance

. mat list `cooksd'

symmetric __000026[1,1]
           y1
y1   .22888166

. matrix `bfd' = diag(bfull)
. matrix `bfdi' = inv(`bfd')
. mat `pctchng'=`bdif'*`bfdi'
. mat `pctchng'=100*`pctchng'
. dis "Percent Change in Coefficients from Full Data Model"
Percent Change in Coefficients from Full Data Model

. mat list `pctchng'

__000025[1,10]
             age     becktota     ndrugfp1     ndrugfp2       ivhx_3         race
y1     8.1298061    3.4170584   -2.2928182   -2.3791833    10.168132    2.3464816

           treat         site     agexsite     racexsit
y1     7.1987089    7.8500058    9.0911235    1.7645622

.
end of do-file
```

In the above listing of the matrix `pctchng' we see that the maximum change is in the coefficient for ivhx_3 at 10.2 percent. The coefficient for age only changes by 8.1 percent. Thus we conclude that the subject with the largest scaled score residual for age is not overly influential.

We proceed to examine the other covariates in a similar manner.

BECKTOTA

`. graph sscr2 becktot`

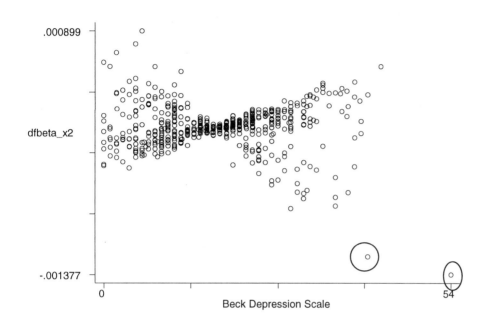

Percent Change in Coefficients from Full Data Model when deleting the subject with the two largest, in absolute value, of the scaled score residual for BECTOTA.

	age	becktota	ndrugfp1	ndrugfp2	ivhx_3	race
y1	.44567852	**33.544342**	-3.2550292	-3.1684959	-10.851063	6.2281053

	treat	site	agexsite	racexsit
y1	-6.5855001	3.0842795	2.0357663	2.2006083

The combined effect is substantial. However as noted in the text the scores of 41 and 54 while large are not that unusual. Thus, we kept the subjects in the analysis.

Over all Change Measure.

We begin by plotting the likelihood displacement diagnostic versus the martingale residuals. We have circled the four subjects with the largest values.

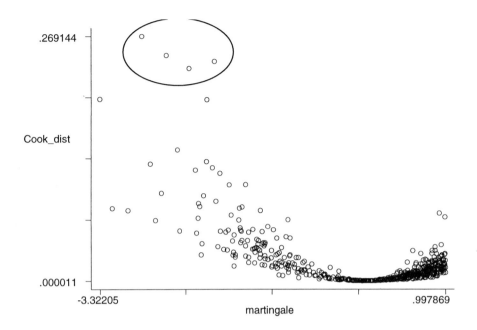

Percent Change in Coefficients from Full Data Model when the subjects with the largest values of the likelihood displacement statistic are deleted.

	age	becktota	ndrugfp1	ndrugfp2	ivhx_3	race
y1	4.8417042	**26.89332**	8.9906514	7.9649555	1.8169366	-.67326876

	treat	site	agexsite	racexsit
y1	13.056903	11.08466	17.376466	**30.734949**

Here we see large changes in the estimated coefficients for BECTOTA, and the RACE by SITE interaction. Next we also delete the two subjects with extreme scaled score residuals for BECTOTA.

Percent Change in Coefficients from Full Data Model

	age	becktota	ndrugfp1	ndrugfp2	ivhx_3	race
y1	5.3461284	**60.837012**	5.7367053	4.8144855	-9.0474778	5.8018432

	treat	site	agexsite	racexsit
y1	6.3175255	14.962076	20.336654	**33.67809**

Here we see quite a large percent change in the coefficient for BECTOTA and a large change for the RACE by SITE interaction. We decide to keep the subjects in as their respective data is not that unusual and they have little or no effect on the coefficient for treatment.

4. *A considerable amount of the material presented in this chapter dealt with the evaluation of fit, and the presentation and interpretation of the fitted model shown in Table 5.11 (and repeated in Table 6.6). Repeat the entire process for the fitted model shown in Table 5.13. This model contains an interaction between AGE and NDRUGFP1 and, as a result, estimation and presentation of hazard ratios for age, controlling for the number of previous drug treatments and for the number of previous drug treatments controlling for age, is a major challenge.*

We leave the evaluation of the fit up to the reader. The steps are the same as that done for the fitted model for the HMO-HIV study and the WHAS model in problem 5, below. We begin by displaying the fit of the new model.

```
. gen agendrug1=age*ndrugfp1
. stcox age beck ndrugfp1 ndrugfp2 ivhx_3 race treat site agendrug1 racesite,
nohr noshow nolog

Cox regression -- Breslow method for ties
No. of subjects =          575                    Number of obs   =        575
No. of failures =          464
Time at risk    =       138900
                                                  LR chi2(10)     =      70.49
Log likelihood  =    -2628.739                    Prob > chi2     =     0.0000
------------------------------------------------------------------------------
         _t |
         _d |      Coef.   Std. Err.      z    P>|z|     [95% Conf. Interval]
------------+-----------------------------------------------------------------
        age |  -.0538471   .0118803    -4.53   0.000    -.077132    -.0305622
   becktota |   .0103485   .0049966     2.07   0.038     .0005554    .0201417
   ndrugfp1 |  -.8377905   .1595497    -5.25   0.000    -1.150502   -.5250788
   ndrugfp2 |   -.229083   .0487574    -4.70   0.000    -.3246458   -.1335202
     ivhx_3 |   .2339795   .1077173     2.17   0.030     .0228575    .4451015
       race |  -.4829436   .1344923    -3.59   0.000    -.7465438   -.2193435
      treat |  -.2223822   .0937517    -2.37   0.018    -.4061322   -.0386321
       site |  -.2781853   .1220049    -2.28   0.023    -.5173106    -.03906
  agendrug1 |   .0072271   .0026092     2.77   0.006     .0021131    .012341
   racesite |   .8968523   .2473127     3.63   0.000     .4121284   1.381576
------------------------------------------------------------------------------
```

The model contains three main effect only covariates: bectota, ivhx_3 and treat. We can estimate the hazard ratios for these variables using the usual method of exponentiating estimated coefficients.

The hazard ratios are in the following table.

Variable	Hazard Ratio	95 % CIE
Beck Score*	1.109	1.006, 1.223
Treatment	0.801	0.667, 0.962
Recent Drug Use	1.264	1.023, 1.561

*: Hazard ratio for a 10 point increase in Beck Score.

Interpretation

(1) Beck Score

Subjects with the 10 point higher score are returning to drug use at a rate that is 11% higher than for subjects at the lower score and it could be as little as a 1 percent increase or as much as a 22 percent increase with 95 percent confidence.

(2) Treatment

Subjects in the longer or extended treatment program are returning to drug use at a rate that is 21% lower than for subjects with the shorter treatment. The 95%confidence interval suggests that the rate could be as much as 33% lower to only 4% lower.

(3) Recent IV Drug use

Subjects with who have a recent history of IV drug use are returning to drug use at a rate that is 26% higher than for subjects who are not recent IV drug users. The confidence interval indicates that the rate could actually be as much as 56% higher or as little as 2 % higher.

(4) Age and Number of Drug Treatments

(4a) Hazard ratio for an increase if one treatment controlling for age.

Step1: Write down the expression for the log-hazard function in the covariates required to estimate the hazard ratio.

$$g(NDRUGTX, AGE, z)$$
$$= \hat{\beta}_1 AGE + \hat{\beta}_2 NDRUGFP1 + \hat{\beta}_3 NDRUGFP2 + \hat{\beta}_4 AGE * NDRUGFP1 + \hat{\beta}z$$

Step2: Write down the equation for the difference of interest in the log hazard function

$$g(NDRUGTX+1, AGE, z) - g(NDRUGTX, AGE, z) =$$
$$\hat{\beta}_1 AGE + \hat{\beta}_2 b + \hat{\beta}_3 bd + \hat{\beta}_4 AGEb + \hat{\beta}z - \left(\hat{\beta}_1 AGE + \hat{\beta}_2 a + \hat{\beta}_3 ac + \hat{\beta}_4 AGEa + \hat{\beta}z\right)$$
$$= \hat{\beta}_2[b-a] + \hat{\beta}_3[bd-ac] + \hat{\beta}_4 AGE[b-a]$$

where

$$a = \frac{10}{NDRUGTX+1}$$
$$b = \frac{10}{NDRUGTX+2}$$
$$c = \ln\left[\frac{NDRUGTX+1}{10}\right] = \ln\left[\frac{1}{a}\right]$$
$$d = \ln\left[\frac{NDRUGTX+2}{10}\right] = \ln\left[\frac{1}{b}\right]$$

To ease the notation let

$$\hat{\xi} = \hat{\beta}_2[b-a] + \hat{\beta}_3[bd-ac] + \hat{\beta}_4 AGE[b-a]$$

The estimate of the hazard ratio is

$$\hat{HR} = \exp(\hat{\xi})$$

Step 3: Calculate the variance of the estimator.

$$\hat{Var}(\hat{\xi}) = [b-a]^2 \hat{Var}(\hat{\beta}_2) + [bd-ac]^2 \hat{Var}(\hat{\beta}_3) + [b-a]^2 AGE^2 \hat{Var}(\hat{\beta}_4)$$
$$+ 2[b-a][bd-ac]\hat{Cov}(\hat{\beta}_2, \hat{\beta}_3) + 2[b-a]^2 AGE \hat{Cov}(\hat{\beta}_2, \hat{\beta}_4) + 2[bd-ac][b-a]AGE \hat{Cov}(\hat{\beta}_3, \hat{\beta}_4)$$

The estimator of the standard error is

$$\hat{SE}(\hat{\xi}) = \sqrt{\hat{Var}(\hat{\xi})}.$$

Step 4: Construct confidence interval

The confidence interval for the hazard ratio has endpoints

$$\exp\left(\hat{\xi} \pm z_{1-\alpha/2}\hat{SE}(\hat{\xi})\right).$$

4(b) Hazard ratio for an increase in AGE given NDRUGTX:

Step 1: Write down the expression for the log-hazard function. The log hazard is

$$g(NDRUGTX, AGE, SITE, z) = \hat{\beta}_1 AGE + \hat{\beta}_2 NDRUGFP1$$
$$+ \hat{\beta}_3 NDRUGFP2 + \hat{\beta}_4 AGE * NDRUGFP1 + \hat{\beta}z$$

Step 2: Write down the equation for the difference of interest in the log hazard function

An increase of w years in age.

$$g(AGE + w, NDRUGTX, z) - g(AGE, NDRUGTX, z) =$$
$$\hat{\beta}_1(AGE + w) + \hat{\beta}_2 a + \hat{\beta}_3 ac + \hat{\beta}_4(AGE + w)a + \hat{\beta}z$$
$$- \left(\hat{\beta}_1 AGE + \hat{\beta}_2 a + \hat{\beta}_3 ac + \hat{\beta}_4 AGE a + \hat{\beta}z\right)$$
$$= \hat{\beta}_1 w + \hat{\beta}_4 wa$$

where a and c are defined above in the expression for hazard ratio for number of drug treatments. Again for ease of notation we let

$$\hat{\xi} = \hat{\beta}_1 w + \hat{\beta}_4 wa$$

and

$$\hat{HR} = e^{\hat{\xi}}$$

Step 3: Calculate variance.

$$\hat{Var}(\hat{\xi}) = w^2 \hat{Var}(\hat{\beta}_1) + w^2 a^2 \hat{Var}(\hat{\beta}_4) + 2w^2 a \hat{Cov}(\hat{\beta}_1, \hat{\beta}_4)$$

and

$$\hat{SE}(\hat{\xi}) = \sqrt{\hat{Var}(\hat{\xi})}$$

Step 4: Construct confidence interval

The endpoints of the confidence interval: are:

$$\exp\left(\hat{\xi} \pm z_{1-\alpha/2}\hat{SE}(\hat{\xi})\right)$$

For the hazard ratios for NDRUGTX given age we use a spreadsheet for easy calculation. We follow this with the calculations for the hazard ratio for age given NDRUGTX and SITE.

NDRUGTX+1 versus NDRUGTX for age=20

NDRUGTX	0	1	2	3	4	5	6	7	8	9	10
AGE	20	20	20	20	20	20	20	20	20	20	20
beta 1	-0.0538	-0.0538	-0.0538	-0.0538	-0.0538	-0.0538	-0.0538	-0.0538	-0.0538	-0.0538	-0.0538
beta 2	-0.8378	-0.8378	-0.8378	-0.8378	-0.8378	-0.8378	-0.8378	-0.8378	-0.8378	-0.8378	-0.8378
beta 3	-0.2291	-0.2291	-0.2291	-0.2291	-0.2291	-0.2291	-0.2291	-0.2291	-0.2291	-0.2291	-0.2291
beta 4	0.0072	0.0072	0.0072	0.0072	0.0072	0.0072	0.0072	0.0072	0.0072	0.0072	0.0072
var 1	0.0001	0.0001	0.0001	0.0001	0.0001	0.0001	0.0001	0.0001	0.0001	0.0001	0.0001
var 2	0.0255	0.0255	0.0255	0.0255	0.0255	0.0255	0.0255	0.0255	0.0255	0.0255	0.0255
var 3	0.0024	0.0024	0.0024	0.0024	0.0024	0.0024	0.0024	0.0024	0.0024	0.0024	0.0024
var 4	0.0000	0.0000	0.0000	0.0000	0.0000	0.0000	0.0000	0.0000	0.0000	0.0000	0.0000
cov 1,2	0.0010	0.0010	0.0010	0.0010	0.0010	0.0010	0.0010	0.0010	0.0010	0.0010	0.0010
cov 1,3	0.0001	0.0001	0.0001	0.0001	0.0001	0.0001	0.0001	0.0001	0.0001	0.0001	0.0001
cov 1.4	0.0000	0.0000	0.0000	0.0000	0.0000	0.0000	0.0000	0.0000	0.0000	0.0000	0.0000
cov 2,3	0.0066	0.0066	0.0066	0.0066	0.0066	0.0066	0.0066	0.0066	0.0066	0.0066	0.0066
cov 2,4	-0.0003	-0.0003	-0.0003	-0.0003	-0.0003	-0.0003	-0.0003	-0.0003	-0.0003	-0.0003	-0.0003
cov 3,4	0.0000	0.0000	0.0000	0.0000	0.0000	0.0000	0.0000	0.0000	0.0000	0.0000	0.0000
a	10.0000	5.0000	3.3333	2.5000	2.0000	1.6667	1.4286	1.2500	1.1111	1.0000	0.9091
b	5.0000	3.3333	2.5000	2.0000	1.6667	1.4286	1.2500	1.1111	1.0000	0.9091	0.8333
c	-2.3026	-1.6094	-1.2040	-0.9163	-0.6931	-0.5108	-0.3567	-0.2231	-0.1054	0.0000	0.0953
d	-1.6094	-1.2040	-0.9163	-0.6931	-0.5108	-0.3567	-0.2231	-0.1054	0.0000	0.0953	0.1823
ln(hazard)	0.0349	0.2313	0.1831	0.1394	0.1085	0.0867	0.0710	0.0592	0.0502	0.0432	0.0376
HR	1.0355	1.2602	1.2009	1.1496	1.1147	1.0906	1.0735	1.0610	1.0515	1.0441	1.0383
var(ln(hazard))	0.0364	0.0035	0.0013	0.0007	0.0004	0.0003	0.0002	0.0001	0.0001	0.0001	0.0000
SE(ln(hazard))	0.1908	0.0593	0.0366	0.0264	0.0202	0.0161	0.0132	0.0110	0.0094	0.0081	0.0070
lower bound	0.7125	1.1220	1.1178	1.0917	1.0713	1.0567	1.0461	1.0383	1.0323	1.0277	1.0240
upper bound	1.5050	1.4155	1.2902	1.2107	1.1598	1.1257	1.1017	1.0842	1.0710	1.0608	1.0527

For 20 years olds subjects with one previous treatment are returning to drug use at a rate that is not significantly different from subjects with no previous treatments. At all other numbers of previous treatments an increase of one treatment significantly increases the rate but the magnitude decreases with the number of treatments. That is the hazard ratio for 2 vs. 1 is 1.26 while 11 vs. 10 it is 1.04.

NDRUGTX+1 versus NDRUGTX for age=25

NDRUGTX	0	1	2	3	4	5	6	7	8	9	10
AGE	25	25	25	25	25	25	25	25	25	25	25
beta 1	-0.0538	-0.0538	-0.0538	-0.0538	-0.0538	-0.0538	-0.0538	-0.0538	-0.0538	-0.0538	-0.0538
beta 2	-0.8378	-0.8378	-0.8378	-0.8378	-0.8378	-0.8378	-0.8378	-0.8378	-0.8378	-0.8378	-0.8378
beta 3	-0.2291	-0.2291	-0.2291	-0.2291	-0.2291	-0.2291	-0.2291	-0.2291	-0.2291	-0.2291	-0.2291
beta 4	0.0072	0.0072	0.0072	0.0072	0.0072	0.0072	0.0072	0.0072	0.0072	0.0072	0.0072
var 1	0.0001	0.0001	0.0001	0.0001	0.0001	0.0001	0.0001	0.0001	0.0001	0.0001	0.0001
var 2	0.0255	0.0255	0.0255	0.0255	0.0255	0.0255	0.0255	0.0255	0.0255	0.0255	0.0255
var 3	0.0024	0.0024	0.0024	0.0024	0.0024	0.0024	0.0024	0.0024	0.0024	0.0024	0.0024
var 4	0.0000	0.0000	0.0000	0.0000	0.0000	0.0000	0.0000	0.0000	0.0000	0.0000	0.0000
cov 1,2	0.0010	0.0010	0.0010	0.0010	0.0010	0.0010	0.0010	0.0010	0.0010	0.0010	0.0010
cov 1,3	0.0001	0.0001	0.0001	0.0001	0.0001	0.0001	0.0001	0.0001	0.0001	0.0001	0.0001
cov 1.4	0.0000	0.0000	0.0000	0.0000	0.0000	0.0000	0.0000	0.0000	0.0000	0.0000	0.0000
cov 2,3	0.0066	0.0066	0.0066	0.0066	0.0066	0.0066	0.0066	0.0066	0.0066	0.0066	0.0066
cov 2,4	-0.0003	-0.0003	-0.0003	-0.0003	-0.0003	-0.0003	-0.0003	-0.0003	-0.0003	-0.0003	-0.0003
cov 3,4	0.0000	0.0000	0.0000	0.0000	0.0000	0.0000	0.0000	0.0000	0.0000	0.0000	0.0000
A	10.0000	5.0000	3.3333	2.5000	2.0000	1.6667	1.4286	1.2500	1.1111	1.0000	0.9091
B	5.0000	3.3333	2.5000	2.0000	1.6667	1.4286	1.2500	1.1111	1.0000	0.9091	0.8333
C	-2.3026	-1.6094	-1.2040	-0.9163	-0.6931	-0.5108	-0.3567	-0.2231	-0.1054	0.0000	0.0953
D	-1.6094	-1.2040	-0.9163	-0.6931	-0.5108	-0.3567	-0.2231	-0.1054	0.0000	0.0953	0.1823
ln(hazard)	-0.1458	0.1711	0.1530	0.1214	0.0965	0.0781	0.0645	0.0542	0.0462	0.0399	0.0348
HR	0.8643	1.1866	1.1653	1.1290	1.1013	1.0813	1.0666	1.0557	1.0473	1.0407	1.0354
var(ln(hazard))	0.0226	0.0017	0.0008	0.0005	0.0003	0.0002	0.0001	0.0001	0.0001	0.0001	0.0000
SE(ln(hazard))	0.1504	0.0412	0.0290	0.0225	0.0179	0.0146	0.0121	0.0102	0.0088	0.0076	0.0067
lower bound	0.6437	1.0946	1.1010	1.0804	1.0633	1.0508	1.0416	1.0347	1.0294	1.0253	1.0220
upper bound	1.1607	1.2863	1.2334	1.1799	1.1407	1.1127	1.0923	1.0771	1.0654	1.0563	1.0490

For 25 years olds, the hazard of returning to drug use is 14% lower, but not significant, for those who had no previous drug treatment compared those who had 1 drug treatment. The point estimate of the hazard ratio for those who had two previous treatments vs. one is 1.19. Thus subjects with two previous treatments are returning to drug use at a rate that is 19% higher than subjects with one previous drug treatment. The effect of an increase of one treatment is a significant increase in the rate of returning to drug use; but as is the case at age 20 the increase diminishes with the number of previous treatments.

NDRUGTX+1 versus NDRUGTX for age=30

NDRUGTX	0	1	2	3	4	5	6	7	8	9	10
AGE	30	30	30	30	30	30	30	30	30	30	30
beta 1	-0.0538	-0.0538	-0.0538	-0.0538	-0.0538	-0.0538	-0.0538	-0.0538	-0.0538	-0.0538	-0.0538
beta 2	-0.8378	-0.8378	-0.8378	-0.8378	-0.8378	-0.8378	-0.8378	-0.8378	-0.8378	-0.8378	-0.8378
beta 3	-0.2291	-0.2291	-0.2291	-0.2291	-0.2291	-0.2291	-0.2291	-0.2291	-0.2291	-0.2291	-0.2291
beta 4	0.0072	0.0072	0.0072	0.0072	0.0072	0.0072	0.0072	0.0072	0.0072	0.0072	0.0072
var 1	0.0001	0.0001	0.0001	0.0001	0.0001	0.0001	0.0001	0.0001	0.0001	0.0001	0.0001
var 2	0.0255	0.0255	0.0255	0.0255	0.0255	0.0255	0.0255	0.0255	0.0255	0.0255	0.0255
var 3	0.0024	0.0024	0.0024	0.0024	0.0024	0.0024	0.0024	0.0024	0.0024	0.0024	0.0024
var 4	0.0000	0.0000	0.0000	0.0000	0.0000	0.0000	0.0000	0.0000	0.0000	0.0000	0.0000
cov 1,2	0.0010	0.0010	0.0010	0.0010	0.0010	0.0010	0.0010	0.0010	0.0010	0.0010	0.0010
cov 1,3	0.0001	0.0001	0.0001	0.0001	0.0001	0.0001	0.0001	0.0001	0.0001	0.0001	0.0001
cov 1,4	0.0000	0.0000	0.0000	0.0000	0.0000	0.0000	0.0000	0.0000	0.0000	0.0000	0.0000
cov 2,3	0.0066	0.0066	0.0066	0.0066	0.0066	0.0066	0.0066	0.0066	0.0066	0.0066	0.0066
cov 2,4	-0.0003	-0.0003	-0.0003	-0.0003	-0.0003	-0.0003	-0.0003	-0.0003	-0.0003	-0.0003	-0.0003
cov 3,4	0.0000	0.0000	0.0000	0.0000	0.0000	0.0000	0.0000	0.0000	0.0000	0.0000	0.0000
a	10.0000	5.0000	3.3333	2.5000	2.0000	1.6667	1.4286	1.2500	1.1111	1.0000	0.9091
b	5.0000	3.3333	2.5000	2.0000	1.6667	1.4286	1.2500	1.1111	1.0000	0.9091	0.8333
c	-2.3026	-1.6094	-1.2040	-0.9163	-0.6931	-0.5108	-0.3567	-0.2231	-0.1054	0.0000	0.0953
d	-1.6094	-1.2040	-0.9163	-0.6931	-0.5108	-0.3567	-0.2231	-0.1054	0.0000	0.0953	0.1823
ln(hazard)	-0.3265	0.1109	0.1229	0.1033	0.0845	0.0695	0.0581	0.0492	0.0422	0.0366	0.0321
HR	0.7215	1.1172	1.1308	1.1088	1.0881	1.0720	1.0598	1.0504	1.0431	1.0373	1.0326
var(ln(hazard))	0.0174	0.0008	0.0006	0.0004	0.0003	0.0002	0.0001	0.0001	0.0001	0.0001	0.0000
SE(ln(hazard))	0.1317	0.0287	0.0240	0.0200	0.0164	0.0136	0.0114	0.0097	0.0084	0.0073	0.0064
lower bound	0.5573	1.0562	1.0787	1.0662	1.0536	1.0438	1.0363	1.0306	1.0261	1.0226	1.0197
upper bound	0.9340	1.1818	1.1853	1.1531	1.1237	1.1010	1.0838	1.0706	1.0603	1.0522	1.0456

The pattern in the estimates of the hazard ratios for 30 years olds is similar to 25 year old subjects; but with slightly smaller effects.

NDRUGTX+1 versus NDRUGTX for age=35

NDRUGTX	0	1	2	3	4	5	6	7	8	9	10
AGE	35	35	35	35	35	35	35	35	35	35	35
beta 1	-0.0538	-0.0538	-0.0538	-0.0538	-0.0538	-0.0538	-0.0538	-0.0538	-0.0538	-0.0538	-0.0538
beta 2	-0.8378	-0.8378	-0.8378	-0.8378	-0.8378	-0.8378	-0.8378	-0.8378	-0.8378	-0.8378	-0.8378
beta 3	-0.2291	-0.2291	-0.2291	-0.2291	-0.2291	-0.2291	-0.2291	-0.2291	-0.2291	-0.2291	-0.2291
beta 4	0.0072	0.0072	0.0072	0.0072	0.0072	0.0072	0.0072	0.0072	0.0072	0.0072	0.0072
var 1	0.0001	0.0001	0.0001	0.0001	0.0001	0.0001	0.0001	0.0001	0.0001	0.0001	0.0001
var 2	0.0255	0.0255	0.0255	0.0255	0.0255	0.0255	0.0255	0.0255	0.0255	0.0255	0.0255
var 3	0.0024	0.0024	0.0024	0.0024	0.0024	0.0024	0.0024	0.0024	0.0024	0.0024	0.0024
var 4	0.0000	0.0000	0.0000	0.0000	0.0000	0.0000	0.0000	0.0000	0.0000	0.0000	0.0000
cov 1,2	0.0010	0.0010	0.0010	0.0010	0.0010	0.0010	0.0010	0.0010	0.0010	0.0010	0.0010
cov 1,3	0.0001	0.0001	0.0001	0.0001	0.0001	0.0001	0.0001	0.0001	0.0001	0.0001	0.0001
cov 1,4	0.0000	0.0000	0.0000	0.0000	0.0000	0.0000	0.0000	0.0000	0.0000	0.0000	0.0000
cov 2,3	0.0066	0.0066	0.0066	0.0066	0.0066	0.0066	0.0066	0.0066	0.0066	0.0066	0.0066
cov 2,4	-0.0003	-0.0003	-0.0003	-0.0003	-0.0003	-0.0003	-0.0003	-0.0003	-0.0003	-0.0003	-0.0003
cov 3,4	0.0000	0.0000	0.0000	0.0000	0.0000	0.0000	0.0000	0.0000	0.0000	0.0000	0.0000
a	10.0000	5.0000	3.3333	2.5000	2.0000	1.6667	1.4286	1.2500	1.1111	1.0000	0.9091
b	5.0000	3.3333	2.5000	2.0000	1.6667	1.4286	1.2500	1.1111	1.0000	0.9091	0.8333
c	-2.3026	-1.6094	-1.2040	-0.9163	-0.6931	-0.5108	-0.3567	-0.2231	-0.1054	0.0000	0.0953
d	-1.6094	-1.2040	-0.9163	-0.6931	-0.5108	-0.3567	-0.2231	-0.1054	0.0000	0.0953	0.1823
ln(hazard)	-0.5071	0.0506	0.0928	0.0852	0.0724	0.0609	0.0516	0.0441	0.0382	0.0333	0.0293
HR	0.6022	1.0519	1.0972	1.0890	1.0751	1.0628	1.0530	1.0451	1.0389	1.0339	1.0298
var(ln(hazard))	0.0206	0.0009	0.0006	0.0004	0.0003	0.0002	0.0001	0.0001	0.0001	0.0001	0.0000
SE(ln(hazard))	0.1435	0.0299	0.0235	0.0195	0.0160	0.0133	0.0112	0.0095	0.0082	0.0071	0.0063
lower bound	0.4546	0.9921	1.0479	1.0482	1.0419	1.0355	1.0302	1.0258	1.0223	1.0195	1.0172
upper bound	0.7978	1.1154	1.1489	1.1313	1.1094	1.0909	1.0763	1.0648	1.0557	1.0484	1.0425

The result for 35 year olds is quite similar to those for 30 year olds.

NDRUGTX+1 versus NDRUGTX for age=40

NDRUGTX	0	1	2	3	4	5	6	7	8	9	10
AGE	40	40	40	40	40	40	40	40	40	40	40
beta 1	-0.0538	-0.0538	-0.0538	-0.0538	-0.0538	-0.0538	-0.0538	-0.0538	-0.0538	-0.0538	-0.0538
beta 2	-0.8378	-0.8378	-0.8378	-0.8378	-0.8378	-0.8378	-0.8378	-0.8378	-0.8378	-0.8378	-0.8378
beta 3	-0.2291	-0.2291	-0.2291	-0.2291	-0.2291	-0.2291	-0.2291	-0.2291	-0.2291	-0.2291	-0.2291
beta 4	0.0072	0.0072	0.0072	0.0072	0.0072	0.0072	0.0072	0.0072	0.0072	0.0072	0.0072
var 1	0.0001	0.0001	0.0001	0.0001	0.0001	0.0001	0.0001	0.0001	0.0001	0.0001	0.0001
var 2	0.0255	0.0255	0.0255	0.0255	0.0255	0.0255	0.0255	0.0255	0.0255	0.0255	0.0255
var 3	0.0024	0.0024	0.0024	0.0024	0.0024	0.0024	0.0024	0.0024	0.0024	0.0024	0.0024
var 4	0.0000	0.0000	0.0000	0.0000	0.0000	0.0000	0.0000	0.0000	0.0000	0.0000	0.0000
cov 1,2	0.0010	0.0010	0.0010	0.0010	0.0010	0.0010	0.0010	0.0010	0.0010	0.0010	0.0010
cov 1,3	0.0001	0.0001	0.0001	0.0001	0.0001	0.0001	0.0001	0.0001	0.0001	0.0001	0.0001
cov 1,4	0.0000	0.0000	0.0000	0.0000	0.0000	0.0000	0.0000	0.0000	0.0000	0.0000	0.0000
cov 2,3	0.0066	0.0066	0.0066	0.0066	0.0066	0.0066	0.0066	0.0066	0.0066	0.0066	0.0066
cov 2,4	-0.0003	-0.0003	-0.0003	-0.0003	-0.0003	-0.0003	-0.0003	-0.0003	-0.0003	-0.0003	-0.0003
cov 3,4	0.0000	0.0000	0.0000	0.0000	0.0000	0.0000	0.0000	0.0000	0.0000	0.0000	0.0000
a	10.0000	5.0000	3.3333	2.5000	2.0000	1.6667	1.4286	1.2500	1.1111	1.0000	0.9091
b	5.0000	3.3333	2.5000	2.0000	1.6667	1.4286	1.2500	1.1111	1.0000	0.9091	0.8333
c	-2.3026	-1.6094	-1.2040	-0.9163	-0.6931	-0.5108	-0.3567	-0.2231	-0.1054	0.0000	0.0953
d	-1.6094	-1.2040	-0.9163	-0.6931	-0.5108	-0.3567	-0.2231	-0.1054	0.0000	0.0953	0.1823
ln(hazard)	-0.6878	-0.0096	0.0627	0.0672	0.0604	0.0523	0.0452	0.0391	0.0341	0.0300	0.0266
HR	0.5027	0.9904	1.0647	1.0695	1.0622	1.0537	1.0462	1.0399	1.0347	1.0305	1.0270
var(ln(hazard))	0.0323	0.0019	0.0008	0.0004	0.0003	0.0002	0.0001	0.0001	0.0001	0.0001	0.0000
SE(ln(hazard))	0.1798	0.0437	0.0276	0.0210	0.0168	0.0137	0.0114	0.0096	0.0083	0.0072	0.0063
lower bound	0.3534	0.9092	1.0086	1.0263	1.0279	1.0258	1.0231	1.0204	1.0181	1.0161	1.0143
upper bound	0.7150	1.0790	1.1238	1.1145	1.0977	1.0824	1.0698	1.0598	1.0517	1.0451	1.0398

The results for 40 year olds show that the hazard ratio for one treatment increase is significant for two or more previous treatments. The increase in the rate, while significant, drops from a only 6 percent increase for three vs. two treatments to a two percent increase for 11 vs. 10 treatments.

104   CHAPTER 6 SOLUTIONS

Below we summarize the results for the hazard ratios for NDRUGTX for given age.

Estimated Hazard Ratio for an Increase of
One Treatment from NDRUGTX

NDRUGTX	AGE				
	20	25	30	35	40
0	1.03	0.86	0.72	0.6	0.5
1	1.26	1.19	1.12	1.05	0.99
2	1.2	1.17	1.13	1.1	1.06
3	1.15	1.13	1.11	1.09	1.07
4	1.11	1.1	1.09	1.075	1.06
5	1.09	1.08	1.07	1.063	1.053
6	1.07	1.07	1.06	1.053	1.046
7	1.06	1.06	1.05	1.045	1.04
8	1.05	1.05	1.043	1.039	1.035
9	1.044	1.04	1.037	1.034	1.03
10	1.038	1.035	1.033	1.03	1.027

We present the graph of the hazard ratio and its confidence interval for up to 10 previous drug treatments.

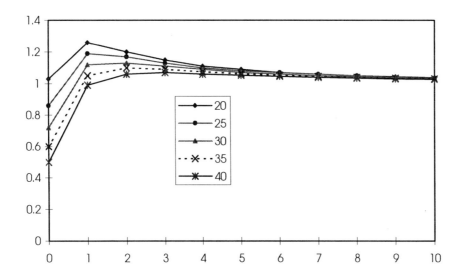

All age groups show a similar pattern. The hazard of returning to drug use is lower for those who had no drug treatments compared to those who had 1 drug treatment and for two or more the estimated hazard ratio is significantly greater than 1.0.

Next we perform the calculations for a 5 year increase in age for ndrugtx = 0, 1, 2, 3, 4 and 5. The respective values of the first fractional polynomial are a = 10, 5, 3.333, 2.5, 2, 1.667.

We use lincom for the calculations as follows

```
. lincom 5*_b[age]+10*5*_b[agxndg1], hr

 ( 1)  5.0 age + 50.0 agxndg1 = 0.0

------------------------------------------------------------------------------
          _t | Haz. Ratio   Std. Err.      z    P>|z|     [95% Conf. Interval]
-------------+----------------------------------------------------------------
         (1) |   1.096494    .105411     0.96   0.338     .9081893    1.323842
------------------------------------------------------------------------------
```

At ndrugtx = 0 we see that there is not a significant effect due to increasing age as the confidence interval includes 1.0, or p = 0.338

```
. lincom 5*_b[age]+5*5*_b[agxndg1], hr

 ( 1)  5.0 age + 25.0 agxndg1 = 0.0

------------------------------------------------------------------------------
          _t | Haz. Ratio   Std. Err.      z    P>|z|     [95% Conf. Interval]
-------------+----------------------------------------------------------------
         (1) |   .9152492   .0422424    -1.92   0.055      .83609    1.001903
------------------------------------------------------------------------------
```

At ndrugtx = 1 we see that there is again no significant effect due to increasing age as the confidence interval includes 1.0, although p = 0.055.

```
. lincom 5*_b[age]+3.333*5*_b[agxndg1], hr

 ( 1)  5.0 age + 16.665 agxndg1 = 0.0

------------------------------------------------------------------------------
          _t | Haz. Ratio   Std. Err.      z    P>|z|     [95% Conf. Interval]
-------------+----------------------------------------------------------------
         (1) |   .8617445   .0350121    -3.66   0.000     .7957832    .9331732
------------------------------------------------------------------------------
```

At ndrugtx = 2 we see that there is a significant protective effect due to increasing age as the confidence interval does not include 1.0. There is an approximate 14 percent reduction in the rate of returning to drug use and it could be between a 7 percent and 20 percent reduction with 95 percent confidence.

The results at ndrugtx = 3, 4 and 5 are similarly protective with the estimate of the effect increasing.

```
. lincom 5*_b[age]+2.5*5*_b[agxndg1], hr

 ( 1)  5.0 age + 12.5 agxndg1 = 0.0

------------------------------------------------------------------------------
          _t | Haz. Ratio   Std. Err.      z    P>|z|     [95% Conf. Interval]
-------------+----------------------------------------------------------------
         (1) |   .8361918   .0351361    -4.26   0.000     .7700859    .9079725
------------------------------------------------------------------------------

. lincom 5*_b[age]+2*5*_b[agxndg1], hr

 ( 1)  5.0 age + 10.0 agxndg1 = 0.0

------------------------------------------------------------------------------
          _t | Haz. Ratio   Std. Err.      z    P>|z|     [95% Conf. Interval]
-------------+----------------------------------------------------------------
         (1) |   .8212195   .0362454    -4.46   0.000     .7531657    .8954223
------------------------------------------------------------------------------

. lincom 5*_b[age]+1.6667*5*_b[agxndg1], hr

 ( 1)  5.0 age + 8.3335 agxndg1 = 0.0

------------------------------------------------------------------------------
          _t | Haz. Ratio   Std. Err.      z    P>|z|     [95% Conf. Interval]
-------------+----------------------------------------------------------------
         (1) |   .8113881   .0373313    -4.54   0.000     .7414222    .8879564
------------------------------------------------------------------------------
```

5. *Repeat the full model evaluation and presentation process using the fitted model developed for the WHAS in problem 3 of Chapter 5.*

Recall that the fitted model contained age, congestive heart failure, shock, MI order, MI type and the interaction between shock and MI order. We begin by refitting the model and saving the various residuals needed to assess the proportional hazards assumption, leverage and influence as well as and baseline survival function.

```
. stcox age chf sho miord mitype01 shoxmiord  ,nohr noshow nolog mgale(M_t)
schoenfeld(sch*) scaledsch(sca*) esr(scr*) bases(S0)

Cox regression -- Breslow method for ties

No. of subjects =         481                     Number of obs   =       481
No. of failures =         249
Time at risk    =      834555
                                                  LR chi2(6)      =    184.61
```

```
Log likelihood  =   -1328.3021                          Prob > chi2      =     0.0000
------------------------------------------------------------------------------
        _t |
        _d |      Coef.   Std. Err.       z     P>|z|     [95% Conf. Interval]
-------------+----------------------------------------------------------------
       age |   .0328211    .005773      5.69   0.000     .0215061    .0441361
       chf |   .5895391   .1422212      4.15   0.000     .3107907    .8682875
       sho |   2.427992   .2677602      9.07   0.000     1.903192    2.952793
     miord |   .4574347   .1402849      3.26   0.001     .1824814    .7323881
   mitype01 |  -.2172837   .1324024     -1.64   0.101    -.4767877    .0422203
  shoxmiord |  -1.120981   .3653705     -3.07   0.002    -1.837094   -.4048682
------------------------------------------------------------------------------
```

(1) Evaluation of the proportional hazards assumption for the covariates.

```
. stphtest, log detail

    Test of proportional hazards assumption

    Time:  Log(t)
    ----------------------------------------------------------------
                 |       rho            chi2       df       Prob>chi2
    -------------+--------------------------------------------------
         age     |    0.03459           0.34        1         0.5620
         chf     |   -0.02153           0.12        1         0.7283
         sho     |   -0.03127           0.24        1         0.6238
         miord   |   -0.12689           3.97        1         0.0462
       mitype01  |    0.22726          13.26        1         0.0003
      shoxmiord  |    0.01416           0.05        1         0.8221
    -------------+--------------------------------------------------
     global test |                     16.87        6         0.0098
    ----------------------------------------------------------------
```

We reject the overall hypothesis on six degrees-of-freedom as $p = 0.0098$. The individual results show highly significant non-proportionality in mitype01 with $p = 0.0003$. There is less striking but still significant non-proportionality in miord with $p = 0.0462$. Similar results are obtained when we use fit the model containing the interactions of the covariates and the log of time.

We show the stphplot for mitype01 below. In the plot the smoothed line does not have slope equal to zero. The plot displays a positive linear relationship, thus supporting a time-varying effect.

```
. stphtest, log plot(mitype01)
```

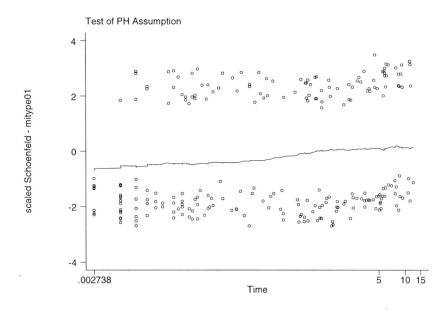

(2) The next step is to plot the scaled score residuals to examine for influence of individual subjects. We examine these in the order that the terms appear in the model.

Age

```
. graph sscr1 age
```

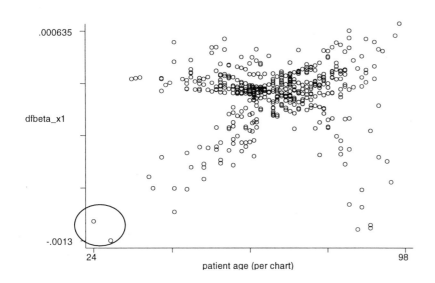

Here we see two values that lie well away from the fan of points. We note that the values for the circled points are similar to two points in the bottom right of the plot. The difference is that those in the lower left lie well away from the next points. The two circled points correspond to the two youngest subjects in the study with age = 24 and 28. The percent changes in the coefficients when these two subjects are deleted are as follows:

```
Percent Change in Coefficients from Full Data Model

           age          chf         sho       miord     mitype01    shoxmiord
y1   7.6090541   -3.6553018   -.1718199    3.4494447    5.0473701    .04314089
```

Note that the maximum percent change is in the age coefficient; but is only 7 percent. Thus these two subjects are not overly influential.

CHF

```
. sort chf
. graph sscr2 , box by(chf)
```

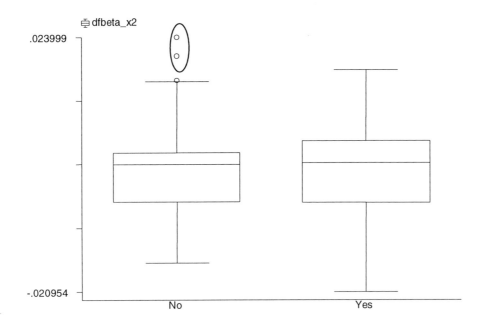

In this plot we see two values among those with chf = 0 that are a bit away from the rest of the values. It is not clear if they lie far enough away to be influential. The percent change in the coefficients when these two subjects are deleted is shown below.

```
Percent Change in Coefficients from Full Data Model
         age         chf         sho       miord     mitype01    shoxmiord
y1   8.3620068  -8.3244578  -.39971983   2.4235197   12.648936   -1.0052197
```

Note that the percent change for all coefficients is less than 15 percent.

SHO

```
. sort sho
. graph sscr3, box by(sho)
```

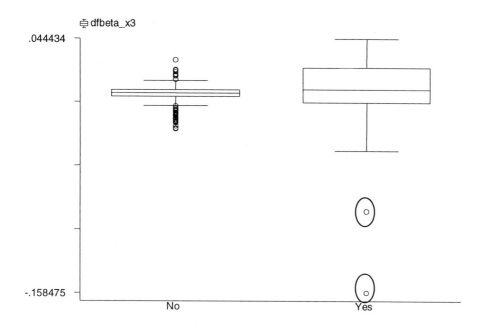

In the plot we see two points among those with sho = 1 that lie well away from the other points. Since shock is involved in an interaction with miord we will wait until we have examined miord and its interaction before deleting any subjects and refitting.

Miord

```
. sort miord
. graph sscr4 , box by(miord)
```

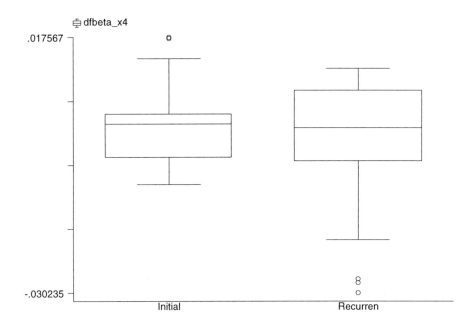

In the plot we see two or three values among subjects with miord=1 that lie well away from the other points. Since miord appears in the model as a main effect we delete these points and refit. Since miord is involved in an interaction with sho we delay deleting and refitting until we examine the interaction.

Mitype01

```
. sort mitype01
. graph sscr5,box by(mitype01)
```

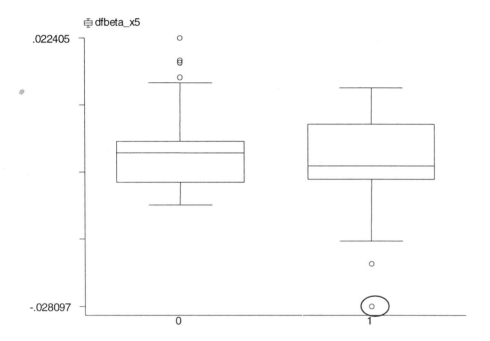

In the plot there is one value that lies away from the rest of the points. We refit the model.

```
Percent Change in Coefficients from Full Data Model
          age         chf         sho       miord     mitype01    shoxmiord
y1  -1.0772537    1.0551758   1.4103092  -.68741946  -16.638019  -19.376491
```

The percent change is greater between 15 and 20 percent for two coefficients.

shoxmiord

```
.gen sho_miord=1*(sho==0)*(miord==0)+2*(sho==0)*(miord==1)/*
*/ +3*(sho==1)*(miord==0)+4*(sho==1)*(miord==1)

. sort sho_miord

. graph sscr6,box by(sho_miord)
```

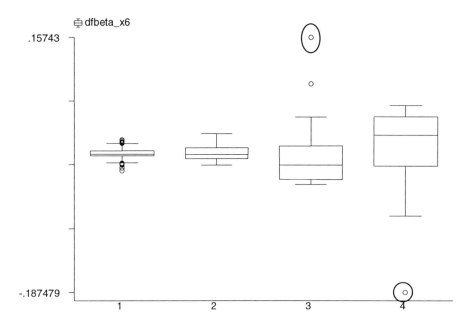

In the above graph we see two points with scaled score residuals that lie well away from the others. The changes in the coefficients when these subjects are deleted are shown below.

```
Percent Change in Coefficients from Full Data Model

              age         chf         sho       miord     mitype01    shoxmiord
y1      3.1457686   .59813766   9.2386332   -.10896804   -5.910017   -2.4279815
```

The percent change is small for all coefficients.

Next we look at the likelihood displacement statistic or Cook's distance graphed versus the martingale residuals.

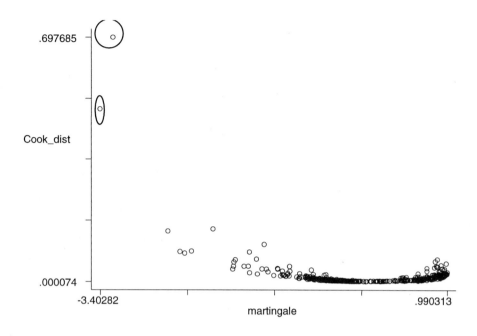

We see two values with large values. Deleting these two subjects and refitting yields the following percent changes in the coefficients

```
Percent Change in Coefficients from Full Data Model

            age         chf         sho        miord    mitype01    shoxmiord
y1    3.1457686   .59813766   9.2386332   -.10896804   -5.910017   -2.4279815
```

Again, the percent change in all the coefficients is quite small.

Since none of the deletions resulted in more than a 20 percent change, we conclude that there are no especially influential subjects. The next step is performing the Grønnesby-Borgan goodness-of-fit test.

To perform the test, we use the STATA do file survfit.do used in earlier problems in this chapter.

(3) Overall test of fit

```
. do survfit M_t 10

. * do file to do Arjas plots
. * and the Gronnesby-Borgan test
. * File will handle stcox and all sreg models
. * Version of 9/12/2001 DWH
. args   M_t G

. capture drop H_t
. capture drop Exp
. capture drop Obs
. capture drop r
. capture drop group
. capture drop zgroup
. capture drop pzgroup

. mat btht=get(_b)

. mat bcoef=btht[1,1..e(df_m)]

. global   S_xvarl : colnames bcoef

. mat drop btht

. mat drop bcoef

. gen H_t= _d -`M_t'

. predict r,xb

. sort r

. gen group=group(`G')

. sort group r

. by group: gen Exp=sum(H_t)

. by group: gen Obs=sum(_d)

. by group: gen zgroup=cond(_n==_N, (Obs-Exp)/sqrt(Exp),.)
(471 missing values generated)

. by group: gen pzgroup=cond(_n==_N,2*(1-norm(abs(zgroup))),.)
(471 missing values generated)

. graph Exp Obs Obs if(_d==1), c(ll) s(ii) by(group) /*
>    */ saving(c:\empty\Arj_$S_E_cmd,replace) b1title("Arjas plot for $S_E_cmd
model")

.
```

```
. more
. quietly xi:$S_E_cmd _t $S_xvar1 i.group, dead(_d)
. lrtest , saving(0)
. quietly xi:$S_E_cmd _t $S_xvar1, dead(_d)
. lrtest
Cox:  likelihood-ratio test                              chi2(9)    =    22.77
                                                         Prob > chi2 =   0.0067

. sort group
. by group : list Obs Exp zgroup pzgroup if _n==_N
```

-> group = 1

	Obs	Exp	zgroup	pzgroup
49.	14	9.736532	1.366346	.1718302

-> group = 2

	Obs	Exp	zgroup	pzgroup
97.	13	14.19304	-.3166773	.7514884

-> group = 3

	Obs	Exp	zgroup	pzgroup
145.	7	18.91562	-2.739721	.0061491

-> group = 4

	Obs	Exp	zgroup	pzgroup
193.	17	21.056	-.8839144	.3767425

-> group = 5

	Obs	Exp	zgroup	pzgroup
241.	21	20.86085	.0304671	.9756945

-> group = 6

	Obs	Exp	zgroup	pzgroup
289.	28	23.58883	.9082401	.3637514

-> group = 7

	Obs	Exp	zgroup	pzgroup
337.	30	26.45456	.6893185	.4906228

-> group = 8

	Obs	Exp	zgroup	pzgroup
385.	36	30.26489	1.042491	.2971843

-> group = 9

	Obs	Exp	zgroup	pzgroup
433.	39	37.05898	.3188485	.7498414

-> group = 10

	Obs	Exp	zgroup	pzgroup
481.	44	46.87071	-.419313	.6749874

```
. exit

end of do-file
```

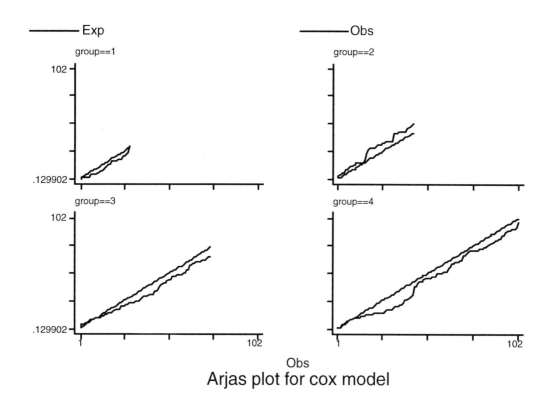

Arjas plot for cox model

Only in the third decile is there a significant difference between the observed and expected number of events. The fit in the remaining deciles seems quite good. We follow this with the Arjas plot by quartile of risk. We could do the plots by decile; but there are two few subjects in each decile for the plot to be informative. We obtain the graph by running survfit.do for 4 groups and suppress the goodness-of-fit test output.

In each plot we see that the cumulative expected is less than the cumulative observed in three quartiles, the polygon falls below the 45-degree line. The observed exceeds the expected in the other quartile. The goodness-of-fit test based on quartiles is also highly significant with $p = 0.006$.

Since we did not find any highly influential subjects we conclude that the lack of fit may be due to non-proportional hazards in miord and mitype01. We explore this further in the next Chapter as well as in Chapter 9. Looking ahead we show that if we stratify the model on mitype01 then we obtain much better adherence to the proportional hazards assumption. Given the lack of model fit it is not a good idea to develop estimates of hazard ratios for this model.

Chapter Seven – Solutions

1. *Problem 3 at the end of Chapter 5 involved finding the best model in the WHAS for LENFOL as survival time and FSTAT as the censoring variable. The fit and adherence to model assumptions was assessed in problem 5 in Chapter 6. In this problem and section we call this model the "WHAS model".*

We begin by displaying the fitted model.

```
.stcox age sho chf miord mitype01 shoxmiord, nohr nolog

Cox regression -- Breslow method for ties

No. of subjects =          481                  Number of obs   =       481
No. of failures =          249
Time at risk    =       834555
                                                LR chi2(6)      =    184.61
Log likelihood  =   -1328.3021                  Prob > chi2     =    0.0000

------------------------------------------------------------------------------
      _t |
      _d |      Coef.   Std. Err.      z    P>|z|     [95% Conf. Interval]
---------+--------------------------------------------------------------------
     age |   .0328211    .005773     5.69   0.000     .0215061    .0441361
     sho |   2.427992   .2677602     9.07   0.000     1.903192    2.952793
     chf |   .5895391   .1422212     4.15   0.000     .3107907    .8682875
   miord |   .4574347   .1402849     3.26   0.001     .1824814    .7323881
mitype01 |  -.2172837   .1324024    -1.64   0.101    -.4767877    .0422203
shoxmiord|  -1.120981   .3653705    -3.07   0.002    -1.837094   -.4048682
------------------------------------------------------------------------------
```

(a) *Treat grouped cohort, YRGRP, as a stratification variable and fit the WHAS model. Compare the estimated coefficients with those from the fit of the WHAS model obtained in problem 3 of Chapter 5. Are there any important differences, i.e. changes greater than 15 percent?*

The model developed previously forces the baseline hazards to be proportional across "study years". If this is not justified, we may use the study year as a stratification variable, whereby each study year would have a separate baseline hazard function.

120 CHAPTER 7 SOLUTIONS

i) Treat grouped cohort, YRGRP, as a stratification variable and fit the WHAS model

```
. stcox age sho chf miord mitype01 shoxmiord, nohr nolog noshow   strata(yrgrp)

Stratified Cox regr. -- Breslow method for ties

No. of subjects =          481                    Number of obs   =       481
No. of failures =          249
Time at risk    =       834555
                                                  LR chi2(8)      =    187.39
Log likelihood  =    -1084.1846                   Prob > chi2     =    0.0000
------------------------------------------------------------------------------
         _t |
         _d |      Coef.   Std. Err.      z    P>|z|     [95% Conf. Interval]
------------+-----------------------------------------------------------------
        age |   .0332147   .0058058     5.72   0.000     .0218356    .0445939
        sho |   2.528599   .2751808     9.19   0.000     1.989255    3.067944
        chf |   .5692496   .1424343     4.00   0.000     .2900835    .8484157
      miord |   .4426207   .1414999     3.13   0.002      .165286    .7199553
    mitype01|  -.1689145   .1350985    -1.25   0.211    -.4337027    .0958737
   shoxmiord|  -1.010381   .3712764    -2.72   0.007    -1.738069   -.2826924
------------------------------------------------------------------------------
                                                         Stratified by yrgrp
```

ii) Comparison of coefficients between the WHAS model and stratified model by year

```
. matrix bfull = get(_b)

. matrix V = get(VCE)

. matrix VI=syminv(V)

. do delta_beta
some output omitted

Percent Change in Coefficients from non-stratified Model

            age         sho         chf       miord     mitype01    shoxmiord
y1    1.1993657   4.1436239  -3.4415888  -3.2385128  -22.260857   -9.866401
```

We note that the percent change in mitype01 is –22% and the other coefficients have not changed in an important manner.

> (b) If the WHAS model examined in problem 5 of Chapter 6 was not proportional in any covariates, examine the effectiveness of using these covariates to define stratification variables. (In this problem do not include YRGRP as a stratification variable.)

i). As we saw in the last Chapter the model was not proportional in mitype01 and miord. We begin by stratifying on mitype01 as non-proportionality was most severe for this variable. Since we are going to have to assess the proportional hazards assumption we save the various residuals.

```
. stcox age chf sho miord shoxmiord ,nohr noshow nolog strata(mitype01)
sch(sch*) scaledsch(sca) esr(scr*) mgale(M_t)

Stratified Cox regr. -- Breslow method for ties

No. of subjects =         481                Number of obs   =         481
No. of failures =         249
Time at risk    =      834555
                                             LR chi2(5)      =      181.66
Log likelihood  =   -1162.0637               Prob > chi2     =      0.0000

------------------------------------------------------------------------------
       _t |
       _d |      Coef.   Std. Err.      z    P>|z|     [95% Conf. Interval]
----------+-------------------------------------------------------------------
      age |   .0337078   .0057883     5.82   0.000     .0223629    .0450527
      chf |   .6063398   .1425757     4.25   0.000     .3268966    .8857829
      sho |   2.364996   .2690691     8.79   0.000     1.83763     2.892362
    miord |   .4361267   .1404759     3.10   0.002     .1607991    .7114543
shoxmiord |  -1.075912   .3668294    -2.93   0.003    -1.794885   -.3569396
------------------------------------------------------------------------------
                                                  Stratified by mitype01
```

```
. stphtest,det log

    Test of proportional hazards assumption

    Time:  Log(t)
    ----------------------------------------------------------------
                |       rho          chi2       df       Prob>chi2
    ------------+---------------------------------------------------
          age   |    0.03063        0.26        1          0.6080
          chf   |   -0.01844        0.09        1          0.7661
          sho   |   -0.03986        0.39        1          0.5316
        miord   |   -0.12160        3.66        1          0.0559
    shoxmiord   |    0.02549        0.16        1          0.6864
    ------------+---------------------------------------------------
    global test |                   4.48        5          0.4826
    ----------------------------------------------------------------
```

We see that the overall test with five degrees-of-freedom is not significant as $p = 0.48$. However, the individual tests demonstrate that the model is marginally non-proportional in miord with $p = 0.056$. In general stratification on mitype01 has improved the adherence to the proportional hazards assumption to the point that we would feel comfortable making inferences on the remaining covariates. We do not show estimates of hazard ratios at this point.

> (c) It is possible that the effect of the covariates in the WHAS model are not constant across cohorts defined by YRGRP. Examine this via inclusion of interactions between YRGRP and model covariates in an extended WHAS model with YRGRP as a stratification variable.

Because of relatively thin data for some model covariates we fit the main effects model and add to it the interactions of the stratification variable and model covariates, one at a time. The result of only one of the separate fits is shown below.

The Wald statistics for the two added interaction variables are only significant for shock. As shown, there is a significant interaction between sho and the design variable for the third grouped cohort. These results show that sho has a significant effect in all grouped cohorts. We use STATA's **lincom** procedure to estimate the effects within strata. It is interesting to note that the estimated hazard ratio in stratum yrgrp=1 is 15.9 while it is 8.39 in stratum yrgrp=2. The fact that the sho by yrgrp=2 interaction term is not significant indicates that these two estimates, while differing by a factor of almost two, are within sampling variation of each other. The estimate in stratum yrgrp=3 is significant and is 4.39. The instability in the estimates is due to the act that the distribution of subjects with shocks is 8, 15 and 15 in the three grouped cohorts.

```
. xi:stcox age i.yrgrp|sho chf miord mitype01, strata(yrgrp) nohr nolog noshow
i.yrgrp           _Iyrgrp_1-3        (naturally coded; _Iyrgrp_1 omitted)
i.yrgrp|sho       _IyrgXsho_#        (coded as above)

Stratified Cox regr. -- Breslow method for ties

No. of subjects =          481                Number of obs   =        481
No. of failures =          249
Time at risk    =       834555
                                              LR chi2(7)      =     184.02
Log likelihood  =    -1085.8707               Prob > chi2     =     0.0000

------------------------------------------------------------------------------
         _t |
         _d |      Coef.   Std. Err.      z    P>|z|     [95% Conf. Interval]
------------+-----------------------------------------------------------------
        age |   .0336217   .0057672     5.83   0.000     .0223182    .0449252
        sho |   2.767037    .489347     5.65   0.000     1.807935     3.72614
 _IyrgXsho_2|  -.639876   .5954323    -1.07   0.283    -1.806902    .5271498
 _IyrgXsho_3|  -1.168703  .5864638    -1.99   0.046    -2.318151   -.0192549
        chf |   .5540808   .1435332     3.86   0.000     .2727609    .8354007
      miord |   .3209111   .1329739     2.41   0.016      .060287    .5815352
   mitype01 |   -.157425   .1358697    -1.16   0.247    -.4237248    .1088747
------------------------------------------------------------------------------
                                                       Stratified by yrgrp
```

```
. lincom _b[sho],hr

 ( 1)  sho = 0.0

------------------------------------------------------------------------------
         _t |  Haz. Ratio   Std. Err.      z    P>|z|     [95% Conf. Interval]
------------+-----------------------------------------------------------------
        (1) |   15.91142    7.786206     5.65   0.000      6.09784    41.51852
------------------------------------------------------------------------------
```

```
. lincom _b[sho] +_b[ _IyrgXsho_2],hr

 ( 1)  sho + _IyrgXsho_2 = 0.0

------------------------------------------------------------------------------
          _t |  Haz. Ratio   Std. Err.       z     P>|z|     [95% Conf. Interval]
-------------+----------------------------------------------------------------
         (1) |    8.391011    2.94376     6.06    0.000      4.218841    16.6892
------------------------------------------------------------------------------

. lincom _b[sho] +_b[ _IyrgXsho_3],hr

 ( 1)  sho + _IyrgXsho_3 = 0.0

------------------------------------------------------------------------------
          _t |  Haz. Ratio   Std. Err.       z     P>|z|     [95% Conf. Interval]
-------------+----------------------------------------------------------------
         (1) |    4.944789   1.623998     4.87    0.000      2.597711   9.412492
------------------------------------------------------------------------------
```

2. *In this chapter we demonstrated, via a time varying dichotomous covariate, that much of the apparent treatment effect in the UIS might be due to differences in length of treatment, LOT. It is quite possible in the WHAS that differences in long term survival in the grouped cohorts, YRGRP, could be due to different lengths of stay in the hospital. Examine this by creating a dichotomous time varying covariate comparing LENSTAY and LENFOL and adding it to the WHAS model.*

In order to create the time-varying covariate we use a function published in the STB 41 called **stegen**. The problem is that this function does not work in STATA 7.0. We note that the STATA 7 manual describes a way to trick STATA 7.0 into creating such a covariate and we illustrate its use at the end of the results for this problem. Since the **stegen** function is so much easier to use we use it here.

```
. version 5
. stset lenfol fstat, id(id)
(you are running stset from Stata version 5)

note:  making entry-time variable t0
       (within id, t0 will be 0 for the 1st observation and the
       lagged value of exit time lenfol thereafter)

    data set name:   C:\WINDOWS\Desktop\whasbk.dta
               id:   id
       entry time:   t0
        exit time:   lenfol
    failure/censor:  fstat
```

```
. stegen out_hos, at(lenstay) from(0) to(1)

number of episode splits : 394
    failure time:   lenfol
      entry time:   t0
   failure/censor:  fstat
              id:   id

. sort id

. list id lenfol lenstay Out_hos      examine the data set changed
          id       lenfol     lenstay     out_hos
  1.       1          1           1          0
  2.       2          1           1          0
  3.       3          1           1          0
  4.       4          1           1          0
  5.       5          2           2          0
  6.       6          2           2          0
  7.       7          2           2          0
  8.       8          3           3          0
  9.       9          3           3          0
 10.      10          3           3          0
 11.      10       5586           3          1
```

Note that subject 10 has had his/her data split into two records corresponding to his/her length of stay in the hospital. The new variable is equal to 0 while he/she is in the hospital, days 1 – 3. The value changes from 0 to 1 when the subject is discharged and remains at 1 until the subject dies after 5586 days. One should examine data for other subjects.

```
862.     473         940          16          1
863.     474          17          17          0
864.     475         779          17          1
865.     475          17          17          0
866.     476        1097          18          1
867.     476          18          18          0
868.     477          20          20          0
869.     477         907          20          1
870.     478         756          22          1
871.     478          22          22          0
872.     479          31          31          0
873.     479         807          31          1
874.     480          39          39          0
875.     481          68          68          0
```

Note: Total number of observation listed is 875 = (481 + 394) for the 481 patients.

Add the time-varying covariate created to the model and assess its significance. Note that we could switch to STATA 7.0 or continue to use version 5. We stay with version 5.

```
. stcox age sho chf miord mitype01 shoxmiord out_hos, nohr nolog
(you are running stcox from Stata version 5)

    failure time:  lenfol
      entry time:  t0
   failure/censor: fstat
              id:  id
(you are running stcox from Stata version 5)

Cox regression -- entry time t0

No. of subjects =          481           Log likelihood = -1322.4761
No. of failures =          249           chi2(7)        =     196.26
Time at risk    =       834555           Prob > chi2    =     0.0000

------------------------------------------------------------------------------
      lenfol |
       fstat |    Coef.    Std. Err.      z     P>|z|    [95% Conf. Interval]
-------------+----------------------------------------------------------------
         age |  .0328879    .0057898    5.68    0.000    .0215401    .0442358
         sho |  2.236546    .2696098    8.30    0.000    1.708121    2.764972
         chf |  .5589113    .142866     3.91    0.000    .278899     .8389235
       miord |  .4527086    .1402567    3.23    0.001    .1778105    .7276066
    mitype01 | -.1750431    .1330158   -1.32    0.188   -.4357492    .085663
   shoxmiord | -.9745547    .3670678   -2.65    0.008   -1.693994   -.2551149
     out_hos | -1.268652    .3844005   -3.30    0.001   -2.022063   -.5152412
------------------------------------------------------------------------------
```

The time-varying covariate we created (out_hos) is significant and negative. This indicates that once a subject is discharged from the hospital their "risk" of dying diminishes greatly. This indicates that patients are not being discharged too early. However, since length of stay has changed in the three grouped cohorts there could be an interaction between out_hos and yrgrp. This model is fit next.

```
. xi: stcox age sho chf miord mitype01 shoxmiord i.yrgrp*out_hos , nohr nolog
i.yrgrp          _Iyrgrp_1-3         (naturally coded; _Iyrgrp_1 omitted)
i.yrgrp*out_hos  _IyrgXout_h_#       (coded as above)

       failure _d:  fstat
  analysis time _t:  (oldt-origin)
           origin:  time oldt0
               id:  id

Cox regression -- Breslow method for ties

No. of subjects =         481                 Number of obs   =        875
No. of failures =         249
Time at risk    =      834555
                                              LR chi2(11)     =     205.21
Log likelihood  =   -1318.0016                Prob > chi2     =     0.0000

------------------------------------------------------------------------------
          _t |
          _d |      Coef.   Std. Err.      z    P>|z|     [95% Conf. Interval]
-------------+----------------------------------------------------------------
         age |   .0335974   .0057988     5.79   0.000     .0222319    .0449629
         sho |   2.318942    .279989     8.28   0.000     1.770173     2.86771
         chf |   .5646963   .1427393     3.96   0.000     .2849324    .8444602
       miord |   .4323744   .1412917     3.06   0.002     .1554477    .7093011
    mitype01 |   -.138938   .1347721    -1.03   0.303    -.4030865    .1252105
   shoxmiord |  -.8547901   .3720006    -2.30   0.022    -1.583898   -.1256824
   _Iyrgrp_2 |  -.6199158   .2754132    -2.25   0.024    -1.159716   -.0801158
   _Iyrgrp_3 |  -.1365116   .2769807    -0.49   0.622    -.6793837    .4063605
     out_hos |  -1.417104   .4242649    -3.34   0.001    -2.248648   -.5855601
 _IyrgXout_~2|   .7074066   .3267547     2.16   0.030     .0669792    1.347834
 _IyrgXout_~3|  -.2062657   .3654949    -0.56   0.573    -.9226226    .5100912
------------------------------------------------------------------------------
```

Here we see that the Wald statistics are significant for one yrgrp coefficient and one interaction of yrgrp with out_hos. At this point we should use **lincom** to estimate the cohort specific effects of out_hos. We leave this to the reader.

Before leaving this problem we present the method for creating the record splitting in STATA 7.

```
. sum  lenstay

    Variable |     Obs        Mean    Std. Dev.       Min        Max
-------------+--------------------------------------------------------
     lenstay |     481     13.85863    8.934296         1         71

. gen enter = 75 - lenstay

. gen exit =75 + (lenfol - lenstay)

. stset exit, enter(time enter) failure(fstat) id(id)

                id:  id
     failure event:  fstat ~= 0 & fstat ~= .
obs. time interval:  (exit[_n-1], exit]
 enter on or after:  time enter
 exit on or before:  failure

------------------------------------------------------------------------
     481  total obs.
       0  exclusions
------------------------------------------------------------------------
     481  obs. remaining, representing
     481  subjects
     249  failures in single failure-per-subject data
  834555  total analysis time at risk, at risk from t =         0
                                  earliest observed entry t =         4
                                       last observed exit t =      5911

. stsplit out_hos, at(0,75)
(394 observations (episodes) created)

. replace out_hos=1 if out_hos==75
(394 real changes made)

. gen t1=_t-_t0

. rename _t oldt
```

```
. stset oldt, origin(oldt0) failure(fstat) id(id)

                id:  id
     failure event:  fstat ~= 0 & fstat ~= .
obs. time interval:  (oldt[_n-1], oldt]
 exit on or before:  failure
    t for analysis:  (time-origin)
            origin:  time oldt0

------------------------------------------------------------------------
      875  total obs.
        0  exclusions
------------------------------------------------------------------------
      875  obs. remaining, representing
      481  subjects
      249  failures in single failure-per-subject data
   834555  total analysis time at risk, at risk from t =         0
                                earliest observed entry t =         0
                                    last observed exit t =      5843

. stcox age sho chf miord mitype01 shoxmiord , nohr nolog

         failure _d:  fstat
   analysis time _t:  (oldt-origin)
             origin:  time oldt0
                 id:  id

Cox regression -- Breslow method for ties

No. of subjects =          481                  Number of obs   =       875
No. of failures =          249
Time at risk    =       834555
                                                LR chi2(6)      =    184.61
Log likelihood  =    -1328.3021                 Prob > chi2     =    0.0000

------------------------------------------------------------------------
         _t |
         _d |      Coef.   Std. Err.      z    P>|z|     [95% Conf. Interval]
------------+-----------------------------------------------------------
        age |   .0328211   .005773     5.69   0.000     .0215061    .0441361
        sho |   2.427992   .2677602    9.07   0.000     1.903192    2.952793
        chf |   .5895391   .1422212    4.15   0.000     .3107907    .8682875
      miord |   .4574347   .1402849    3.26   0.001     .1824814    .7323881
   mitype01 |  -.2172837   .1324024   -1.64   0.101    -.4767877    .0422203
  shoxmiord |  -1.120981   .3653705   -3.07   0.002    -1.837094   -.4048682
------------------------------------------------------------------------
```

```
. stcox age sho chf miord mitype01 shoxmiord out_hos , nohr nolog

        failure _d:  fstat
   analysis time _t:  (oldt-origin)
            origin:  time oldt0
                id:  id

Cox regression -- Breslow method for ties

No. of subjects =          481                Number of obs   =         875
No. of failures =          249
Time at risk    =       834555
                                              LR chi2(7)      =      196.26
Log likelihood  =    -1322.4761               Prob > chi2     =      0.0000

------------------------------------------------------------------------------
         _t |
         _d |      Coef.   Std. Err.      z    P>|z|     [95% Conf. Interval]
------------+-----------------------------------------------------------------
        age |   .0328879   .0057898     5.68   0.000     .0215401    .0442358
        sho |   2.236546   .2696098     8.30   0.000     1.708121    2.764972
        chf |   .5589113    .142866     3.91   0.000      .278899    .8389235
      miord |   .4527086   .1402567     3.23   0.001     .1778105    .7276066
   mitype01 |  -.1750431   .1330158    -1.32   0.188    -.4357492     .085663
  shoxmiord |  -.9745547   .3670678    -2.65   0.008    -1.693994   -.2551149
    out_hos |  -1.268652   .3844005    -3.30   0.001    -2.022063   -.5152412
------------------------------------------------------------------------------
```

Note that the results of fitting the base model and the one containing out_hos are the same as previous fits of these two models.

3. *An alternative approach to the analysis of long term survival in the WHAS is to study the survival experience of patients post-hospital discharge using LENFOL as survival time. Restrict the analysis to patients discharged alive, but account for differing lengths of stay by defining LENSTAY as the delayed entry time. That is, use counting process-type input of the form (LENSTAY, LENFOL, FSTAT). Note that subjects with LENSTAY = LENFOL must be excluded as 0 is not an allowable value for survival time in any software package. Alternatively one could add a small constant (less than 1) to the value of LENFOL for these subjects. Examine the effect on the WHAS model of this alternative method of assessing long term survival in the WHAS. The analysis in this problem could be performed at several levels. The simplest analysis is just a refit of the WHAS model from problem 3 of Chapter 4, but with delayed entry. The most elaborate analysis involves a complete repeat of all steps performed in the model building and assessment in Chapters 5 and 6.*

For simplicity, we will just refit of the WHAS model from problem 3 of Chapter 4, but with delayed entry.

130 *CHAPTER 7 SOLUTIONS*

(i) Subjects with LENSTAY = LENFOL are excluded as 0 is not an allowable value for survival time.

```
. drop if dstat==1      Restrict the analysis to patients discharged alive
(82 observations deleted)
. drop if lenfol== lenstay
(5 observations deleted)

. stset lenfol, failure(fstat) time0(lenstay)

     failure event:  fstat ~= 0 & fstat ~= .
obs. time interval:  (lenstay, lenfol]
 exit on or before:  failure
------------------------------------------------------------------------
       394  total obs.
         0  exclusions
------------------------------------------------------------------------
       394  obs. remaining, representing
       163  failures in single record/single failure data
    827889  total analysis time at risk, at risk from t =         0
                              earliest observed entry t =         1
                                   last observed exit t =      5843

. stcox age chf sho miord mitype01 shoxmiord ,nohr noshow nolog

Cox regression -- Breslow method for ties

No. of subjects =         394                Number of obs   =       394
No. of failures =         163
Time at risk    =      827889
                                             LR chi2(6)      =     66.96
Log likelihood  =   -855.25984               Prob > chi2     =    0.0000

------------------------------------------------------------------------
          _t |
          _d |     Coef.   Std. Err.      z    P>|z|   [95% Conf. Interval]
-------------+----------------------------------------------------------
         age |   .0312574    .007001    4.46   0.000    .0175358    .044979
         chf |   .6479646   .1681027    3.85   0.000    .3184893   .9774399
         sho |   3.518022    1.10715    3.18   0.001    1.348048   5.687996
      miord1 |   .3839774   .1635329    2.35   0.019    .0634588    .704496
    mitype01 |   .0649864   .1618148    0.40   0.688   -.2521648   .3821376
   shoxmiord |  -2.916269   1.255219   -2.32   0.020   -5.376454  -.4560848
------------------------------------------------------------------------
```

When we compare the model fit using delayed entry to the original model we see that the effect of sho has increased dramatically and the effect of mitype01 is no longer significant. The coefficients for the other variables are more or less the same.

(Now we examine what happens if we allow the five subjects with lensty = lenfol into the analysis by adding a small constant (less than 1) to the value of lenfol for these subjects.

First clear the existing data and start with the original data set of size 481.

```
. drop if dstat==1
(82 observations deleted)
. set seed 05734
. replace lenfol= lenfol+uniform() if lenstay== lenfol
(5 real changes made)
. stset lenfol, failure(fstat) time0(lenstay)

     failure event:  fstat ~= 0 & fstat ~= .
obs. time interval:  (lenstay, lenfol]
 exit on or before:  failure
------------------------------------------------------------------------
      399  total obs.
        0  exclusions
------------------------------------------------------------------------
      399  obs. remaining, representing
      167  failures in single record/single failure data
 827892.1  total analysis time at risk, at risk from t =         0
                                earliest observed entry t =         1
                                     last observed exit t =      5843


. stcox age chf sho miord mitype01 shoxmiord ,nohr noshow nolog

Cox regression -- Breslow method for ties

No. of subjects =         399                 Number of obs   =       399
No. of failures =         167
Time at risk    =   827892.0934
                                              LR chi2(6)      =     69.34
Log likelihood  =    -871.29009               Prob > chi2     =    0.0000

------------------------------------------------------------------------
     _t |
     _d |     Coef.   Std. Err.      z    P>|z|    [95% Conf. Interval]
--------+---------------------------------------------------------------
    age |   .0307467   .0069048    4.45   0.000    .0172136    .0442798
    chf |   .6657169   .1665014    4.00   0.000    .3393802    .9920537
    sho |   3.338739   1.091824    3.06   0.002    1.198804    5.478674
  miord1|   .4082381   .1611352    2.53   0.011    .0924189    .7240573
mitype01|   .0627794   .1600018    0.39   0.695   -.2508183    .3763771
shoxmiord| -2.785356   1.241711   -2.24   0.025   -5.219065   -.3516462
------------------------------------------------------------------------
```

By using the delta_beta.do program and using as the "full" model the one based on 394 subjects we obtain the following percent change in coefficients.

```
Percent Change in Coefficients from Full Data Model
           age         chf         sho       miord1    mitype01   shoxmiord
y1   1.6608787  -2.6666483   5.3697706   -5.9427756   3.5154475   4.7000746
```

Thus we see that "jittering" lenfol for the five subjects with lenfol = lenstay has not altered the model appreciably.

132 CHAPTER 7 SOLUTIONS

4. *The survival time variable LENFOL is calculated as the days between the hospital admission date, HOSPDAT, and the date of the last follow-up, FOLDAT. Follow-up of patients is not done continuously but at various intervals. Thus it is possible that a reported number of days of follow-up could be inaccurate. In order to explore this, create a new variable MNTHFOL = LENFOL/30.416667. Use MNTHFOL to create a discrete, in multiples 24 months, (e.g., 2 year intervals). For purposes of this problem, restrict analysis to those subjects with at most 120 months of follow-up.*

   (a) *Use the method for fitting interval censored data presented in Section 7.5, fit the WHAS model from problem 3 in Chapter 5. Compare the results of the two fitted models. Is this comparison helpful or not in evaluating the described grouping strategy as a method for dealing with imprecisely measured survival times?*

(i) Method for fitting interval censored (time unit = 2 year interval)

```
. gen mnthfol= lenfoll/30.416667    Change a time scale from day to month

. gen yrfol=recode( mnthfol, 24, 48, 72, 96, 120, 144, 168, 192)/12
                                    Change a time scale from day to month

. tab yrfol fstat
       | follow-up status as
       |    of last dat
 yrfol |    Alive      Dead |     Total
-------+--------------------+----------
     2 |        1       169 |       170
     4 |       46        27 |        73
     6 |       45        26 |        71
     8 |       43         8 |        51
    10 |       38        12 |        50
    12 |        1         6 |         7
    14 |       32         1 |        33
    16 |       26         0 |        26
-------+--------------------+----------
 Total |      232       249 |       481

. drop if yrfol>10    Restrict to subjects with at most 120 months of follow-up.
(66 observations deleted)

. generate interval = yrfol/2    2 year intervals
```

```
. tab interval fstat

           |  follow-up status as
           |      of last dat
  interval |     Alive       Dead |     Total
-----------+----------------------+----------
         1 |         1        169 |       170
         2 |        46         27 |        73
         3 |        45         26 |        71
         4 |        43          8 |        51
         5 |        38         12 |        50
-----------+----------------------+----------
     Total |       173        242 |       415


. sort id
. expand interval    expand replaces each observation in the current dataset
(568 observations created)
. sort id

. quietly by id:generate intnew=_n         nth copy of the observation for id
. quietly by id:generate z=cond(_n==_N,fstat,0)
```

CHAPTER 7 SOLUTIONS

```
. list id yrfol interval intnew fstat z

            id    yrfol   interval    intnew     fstat      z
  1.         1        2          1         1      Dead      1
  2.         2        2          1         1      Dead      1
  3.         3        2          1         1      Dead      1
  4.         4        2          1         1      Dead      1
  5.         5        2          1         1      Dead      1
  6.         6        2          1         1      Dead      1
  7.         7        2          1         1      Dead      1
  8.         8        2          1         1      Dead      1
  9.         9        2          1         1      Dead      1
 10.        11        2          1         1      Dead      1
 11.        12        2          1         1      Dead      1
 12.        13        2          1         1      Dead      1
 13.        14        2          1         1      Dead      1
 14.        15        6          3         1      Dead      0
 15.        15        6          3         2      Dead      0
 16.        15        6          3         3      Dead      1

deleted output

967.       472        4          2         1     Alive      0
968.       472        4          2         2     Alive      0
969.       473        4          2         1      Dead      0
970.       473        4          2         2      Dead      1
971.       474        2          1         1      Dead      1
972.       475        4          2         1     Alive      0
973.       475        4          2         2     Alive      0
974.       476        4          2         1     Alive      0
975.       476        4          2         2     Alive      0
976.       477        4          2         1     Alive      0
977.       477        4          2         2     Alive      0
978.       478        4          2         1     Alive      0
979.       478        4          2         2     Alive      0
980.       479        4          2         1     Alive      0
981.       479        4          2         2     Alive      0
982.       480        2          1         1      Dead      1
983.       481        2          1         1      Dead      1


. tab intnew, gen(int)      Create int1-int5 variables

    intnew |      Freq.     Percent        Cum.
-----------+-----------------------------------
         1 |        415       42.22       42.22
         2 |        245       24.92       67.14
         3 |        172       17.50       84.64
         4 |        101       10.27       94.91
         5 |         50        5.09      100.00
-----------+-----------------------------------
     Total |        983      100.00
```

```
.glm z age sho chf miord mitype01 shoxmiord int1 int2 int3 int4 int5, nocons nolog
family(binomial) link(clog)

Generalized linear models                       No. of obs      =         983
Optimization     : ML: Newton-Raphson           Residual df     =         972
                                                Scale param     =           1
Deviance         =    842.4312962               (1/df) Deviance =    .8666989
Pearson          =     1004.88652               (1/df) Pearson  =    1.033834

Variance function: V(u) = u*(1-u)               [Bernoulli]
Link function    : g(u) = ln(-ln(1-u))          [Complementary log-log]
Standard errors  : OIM

Log likelihood   = -421.2156481                 AIC             =    .8793808
BIC              =    766.6345959

------------------------------------------------------------------------------
           z |      Coef.   Std. Err.      z    P>|z|     [95% Conf. Interval]
-------------+----------------------------------------------------------------
         age |    .029556    .0060328     4.90   0.000     .017732    .0413799
         sho |   4.150972    53.09292     0.08   0.938    -99.90924   108.2112
         chf |   .7135244    .1527031     4.67   0.000     .4142318   1.012817
       miord |    .443752    .1468143     3.02   0.003     .1560013    .7315028
    mitype01 |  -.3255753    .1430951    -2.28   0.023    -.6060365    -.045114
   shoxmiord |  -3.203736    53.09414    -0.06   0.952    -107.2663   100.8589
        int1 |  -3.223617     .43051     -7.49   0.000    -4.067401   -2.379832
        int2 |  -4.427201    .456033     -9.71   0.000    -5.321009   -3.533392
        int3 |  -4.054578   .4489237     -9.03   0.000    -4.934453   -3.174704
        int4 |   -4.71154   .5352778     -8.80   0.000    -5.760665   -3.662415
        int5 |  -3.241684   .4864517     -6.66   0.000    -4.195112   -2.288257
------------------------------------------------------------------------------
```

Note that the coefficient for sho and shoxmiord are quite large with an extremely large standard errors. This indicates that there may be some potentially severe numerical problems in the expanded data set.

In order to better compare the approaches we fit the model excluding the interaction.

```
. glm z age sho chf miord mitype01  int1 int2 int3 int4 int5, nocons nolog
family(binomial) link(clog)

Generalized linear models                        No. of obs      =        983
Optimization     : ML: Newton-Raphson            Residual df     =        973
                                                 Scale param     =          1
Deviance         =  848.9517814                  (1/df) Deviance =   .8725095
Pearson          =  1015.613365                  (1/df) Pearson  =   1.043796

Variance function: V(u) = u*(1-u)                [Bernoulli]
Link function    : g(u) = ln(-ln(1-u))           [Complementary log-log]
Standard errors  : OIM

Log likelihood   = -424.4758907                  AIC             =   .8839794
BIC              =  780.0456902

------------------------------------------------------------------------------
           z |      Coef.   Std. Err.      z    P>|z|     [95% Conf. Interval]
-------------+----------------------------------------------------------------
         age |   .0296796   .0059448     4.99   0.000      .018028    .0413311
         sho |   1.496223   .2742516     5.46   0.000     .9586997    2.033746
         chf |   .6908133    .150293     4.60   0.000     .3962443    .9853822
       miord |   .3627766   .1430391     2.54   0.011     .0824251    .6431282
    mitype01 |  -.3576578   .1419828    -2.52   0.012     -.635939   -.0793767
        int1 |  -3.165354   .4240964    -7.46   0.000    -3.996568   -2.334141
        int2 |  -4.393043   .4506223    -9.75   0.000    -5.276246   -3.509839
        int3 |  -4.029377   .4440521    -9.07   0.000    -4.899704   -3.159051
        int4 |  -4.664382   .5303977    -8.79   0.000    -5.703943   -3.624822
        int5 |  -3.204109   .4819814    -6.65   0.000    -4.148775   -2.259443
------------------------------------------------------------------------------
```

Here we see that all the estimated coefficients have reasonable values.

Now we fit the model to the actual length of follow-up.

```
. clear

. use whas

Note: since the proportional hazards partial likelihood only depends on the rank of the
observations then we do not need to convert to days, only drop observations with
mnthfol>120.

. drop if  mnthfol>120  Restrict to subjects with at most 120 months of follow-up
(66 observations deleted)

. stcox age sho chf miord mitype01, nohr nolog noshow

Cox regression -- Breslow method for ties

No. of subjects =         415                    Number of obs   =        415
No. of failures =         242
Time at risk    =      507287
                                                 LR chi2(5)      =     160.99
Log likelihood  =   -1243.5232                   Prob > chi2     =     0.0000

------------------------------------------------------------------------------
        _t |
        _d |      Coef.   Std. Err.      z    P>|z|     [95% Conf. Interval]
-----------+------------------------------------------------------------------
       age |    .028745    .005656     5.08   0.000     .0176595    .0398305
       sho |   1.631194     .20432     7.98   0.000     1.230734    2.031653
       chf |   .6907162   .1481979     4.66   0.000     .4002537    .9811788
     miord1|   .2719737   .1352398     2.01   0.044     .0069086    .5370389
   mitype01| -.3554101   .1338474    -2.66   0.008    -.6177462   -.0930741
------------------------------------------------------------------------------
```

(iii) Compare the results of the two fitted models.

The estimates of the coefficients are quite similar.

> (b) Use the fitted model in 4.1 and present covariate adjusted survivorship functions comparing the survivorship experience of patients with and without complications due to cardiogenic shock.

Due to the problems with the shoxmiord interaction we do this problem using the main effects only model. We estimate the survival functions on the original data, not the expanded data used to estimate the coefficients.

```
. gen mnthfol= lenfol/30.416667    Change a time scale from day to month

. gen yrfol=recode( mnthfol, 24, 48, 72, 96, 120, 144, 168, 192)/12
                                   Change a time scale from day to month

. tab yrfol fstat
         | follow-up status as
         |    of last dat
   yrfol |    Alive        Dead |    Total
---------+----------------------+---------
       2 |        1         169 |      170
       4 |       46          27 |       73
       6 |       45          26 |       71
       8 |       43           8 |       51
      10 |       38          12 |       50
      12 |        1           6 |        7
      14 |       32           1 |       33
      16 |       26           0 |       26
---------+----------------------+---------
   Total |      232         249 |      481

. drop if yrfol>10    Restrict to subjects with at most 120 months of follow-up.
(66 observations deleted)

. generate interval = yrfol/2    2 year intervals

. tab interval

   interval |     Freq.       Percent         Cum.
-----------+-----------------------------------------
          1 |       170         40.96        40.96
          2 |        73         17.59        58.55
          3 |        71         17.11        75.66
          4 |        51         12.29        87.95
          5 |        50         12.05       100.00
-----------+-----------------------------------------
     Total |       415        100.00


. scalar c1 = exp(-exp(-3.165354))

  scalar c2=c1*exp(-exp(-4.393043))

. scalar c3=c2*exp(-exp(-4.029377))

. scalar c4=c3*exp(-exp(-4.664382))

. scalar c5=c4*exp(-exp(-3.204109))

. dis c1 c2 c3 c4 c5
.95867879   .94689956   .93020744   .92148136   .88482288
```

```
. gen S0 = c1*(interval==1) + c2*(interval==2) + c3*(interval==3) + c4*(interval==4) +
  c5*(interval==5)

. generate rm=.0296796 *age+.6908133*chf+.3627766* miord-.3576578* mitype01
    Coefficients are from 4.a.(ii)

. sum rm,det
```

```
                                  rm
-------------------------------------------------------------
        Percentiles      Smallest
 1%       1.007604       .7123104
 5%       1.245041       .8295262
10%        1.40006       .8592058     Obs                 415
25%       1.815574        .918565     Sum of Wgt.         415

50%       2.260768                    Mean           2.304085
                          Largest     Std. Dev.      .6513134
75%       2.832863       3.635715
90%       3.160841       3.665395     Variance       .4242092
95%       3.368599       3.695074     Skewness       .0109376
99%       3.606035       3.724754     Kurtosis       2.188813

. gen s_sho_0=S0^exp(2.260768)     Median=2.260768
. gen s_sho_1=S0^exp(2.26518+1.496223)   Coefficient for sho=1.513545
```

. graph   s_sho_0 s_sho_1 yrfol, yscale(0,1) xscale(0,10) xlabel(0,2,4,6,8,10) c(JJ)

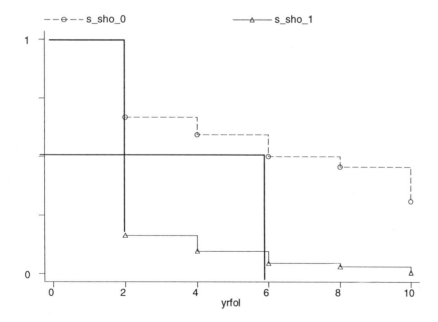

(c)   Use the graphs in 4.2 to estimate the covariate adjusted median survival time for the two levels of cardiogenic shock.

Median survival time for No cardiogenic shock group: 6 years
Median survival time for cardiogenic shock group : 2 years

(d)   Describe a strategy for including grouped cohort, YRGRP, as a stratification variable in the interval censored data analysis in 4.2. If possible implement this strategy.

The idea is essentially the same as fitting a stratified logistic regression model, e.g., see Hosmer & Lemeshow, 2000, Chapter 7. This can be done by extending the previous model to include interactions between YRGRP and the interval dummy variables. This can be done using the same command as used to fit the model shown in 4.1 but including the interaction terms. Given the thin data we can expect numerical problems.

Chapter Eight – Solutions

The first three problems involve simple two variable exponential and Weibull regression models fit to data from the UIS. The goal of these problems is to estimate and compare time and hazard ratios in a simple setting. For this reason, important model building details, such as checking the scale of age, are skipped.

1. Fit an exponential regression model containing AGE and TREAT to the UIS data.

Note: We may use one of the following three different commands (**stereg**, **ereg**, **streg**) for exponential regression model. We will get exactly the same computer output.

```
. streg treat age,  dist(exponential) time

       failure _d:  censor
   analysis time _t:  time

Exponential regression -- log expected-time form

No. of subjects =            623                 Number of obs   =         623
No. of failures =            504
Time at risk    =         146816
                                                 LR chi2(2)      =       10.78
Log likelihood  =     -1021.7791                 Prob > chi2     =      0.0046
------------------------------------------------------------------------------
         _t |      Coef.   Std. Err.      z    P>|z|     [95% Conf. Interval]
------------+----------------------------------------------------------------
      treat |   .2351175   .0891587     2.64   0.008     .0603696    .4098653
        age |   .0140258   .0071984     1.95   0.051    -.0000829    .0281345
      _cons |   5.099416   .2399786    21.25   0.000     4.629067    5.569766
------------------------------------------------------------------------------
```

time option

time specifies that the model is to be estimated in the accelerated failure-time metric rather than in the log relative-hazard metric. This option is only valid for the exponential and Weibull models since they have both a proportional hazards and a accelerated failure-time parameterization. For these two models, in the log relative-hazard metric, estimates of the coefficients are produced and in the accelerated failure-time metric, estimates of coefficients and shape parameter are produced.

(a) *Using the fitted model compute point and 95 percent confidence interval estimates of the time ratio for treatment and for a 10-year increase in age. Interpret these estimates within the context of the UIS.*

```
. lincom _b[treat], eform

 ( 1)  [_t]treat = 0.0

------------------------------------------------------------------------
      _t |    exp(b)   Std. Err.      z    P>|z|    [95% Conf. Interval]
---------+--------------------------------------------------------------
     (1) |  1.265057   .1127909    2.64   0.008    1.062229    1.506615
------------------------------------------------------------------------
```

```
. lincom 10* _b[age], eform
 ( 1)  10.0 [_t]age = 0.0

------------------------------------------------------------------------
      _t |    exp(b)   Std. Err.      z    P>|z|    [95% Conf. Interval]
---------+--------------------------------------------------------------
     (1) |  1.150571   .0828232    1.95   0.051     .9991715   1.324911
------------------------------------------------------------------------
```

Interpretation of estimated time ratios

Time to return to drug for the subjects on the longer treatment is 26.5 percent longer than for subjects on the shorter treatment and it could be as little as 6 percent or a much as 51 percent loner with 95 percent confidence.

The estimated time ratio for age is 1.15. This implies that each 10 year increase in age results in an estimate of 15 percent longer time to return to drug use. Since p = 0.051 the beneficial effect of increasing age is just barely not significant at the 5 percent level. The confidence interval suggests that the treatment effect could result in returning to drug use 1 percent sooner or up to 32 percent longer.

(b) *Using the fitted model compute point and 95 percent confidence interval estimates of the hazard ratio for treatment and for a 10-year change in age. Interpret these estimates within the context of the UIS.*

```
. streg age treat, dist(exp)   nolog noshow nohr
Exponential regression -- log relative-hazard form

No. of subjects =         623               Number of obs    =       623
No. of failures =         504
Time at risk    =      146816
                                            LR chi2(2)       =     10.78
Log likelihood  =   -1021.7791              Prob > chi2      =    0.0046
------------------------------------------------------------------------------
         _t |      Coef.   Std. Err.      z    P>|z|     [95% Conf. Interval]
------------+-----------------------------------------------------------------
        age |  -.0140258   .0071984    -1.95   0.051    -.0281345    .0000829
      treat |  -.2351175   .0891587    -2.64   0.008    -.4098653   -.0603696
      _cons |  -5.099416   .2399786   -21.25   0.000    -5.569766   -4.629067
------------------------------------------------------------------------------

. lincom _b[treat], hr
 ( 1)  [_t]treat = 0.0
------------------------------------------------------------------------------
         _t | Haz. Ratio   Std. Err.      z    P>|z|     [95% Conf. Interval]
------------+-----------------------------------------------------------------
        (1) |    .790478    .070478    -2.64   0.008     .6637396    .9414165
------------------------------------------------------------------------------

. lincom 10* _b[age], hr
 ( 1)  10.0 [_t]age = 0.0
------------------------------------------------------------------------------
         _t | Haz. Ratio   Std. Err.      z    P>|z|     [95% Conf. Interval]
------------+-----------------------------------------------------------------
        (1) |   .8691339   .0625641    -1.95   0.051     .7547679    1.000829
------------------------------------------------------------------------------
```

Subjects on the longer treatment are returning to drug use at a rate that is 21% less than subjects on the shorter treatment.

The estimate of the hazard ratio for age indicates that the rate of returning to drug use decreases by 13 percent for each 10 year increase in age.

(c) Compare the time ratio and hazard ratio estimates computed in problems 1(a) and 1(b). In particular, which estimate, time or hazard ratio, do you think would be more easily understood by non-statistically trained study staff?

The time ratios, since they apply to the more familiar time metric may be more easily understood than hazard ratios, which compare rates, especially by investigators who are new to regression models for survival data.

(d) Why can the fitted exponential regression model be used to compute both time and hazard ratio estimates?

The exponential and Weibull distributions are the only accelerated failure time models that are also proportional hazards models.

2. Repeat problem 1 using the Weibull regression model.

(a) Using the fitted model compute point and 95 percent confidence interval estimates of the time ratio for treatment and for a 10-year increase in age. Interpret these estimates within the context of the UIS.

```
. streg treat age, nolog dist(weibull) time

         failure _d:  censor
   analysis time _t:  time

Weibull regression -- accelerated failure-time form

No. of subjects =          623                  Number of obs   =        623
No. of failures =          504
Time at risk    =       146816
                                                LR chi2(2)      =       9.63
Log likelihood  =   -1017.0837                  Prob > chi2     =     0.0081
------------------------------------------------------------------------------
          _t |      Coef.   Std. Err.      z    P>|z|     [95% Conf. Interval]
-------------+----------------------------------------------------------------
       treat |   .2479208   .0996765     2.49   0.013     .0525586    .4432831
         age |   .0150457    .008056     1.87   0.062    -.0007438    .0308352
       _cons |   5.046659   .2691065    18.75   0.000     4.51922     5.574098
-------------+----------------------------------------------------------------
       /ln_p |  -.1105534   .0370026    -2.99   0.003    -.1830772   -.0380296
-------------+----------------------------------------------------------------
           p |   .8953385   .0331299                      .8327038    .9626844
         1/p |   1.116896   .0413281                      1.038762    1.200907
------------------------------------------------------------------------------
```

We note that the parametrization presented in the above STATA output is slightly different from what is shown in the text. In version 7 STATA has chosen to present output in terms of λ rather that σ. In the above table $\hat{\lambda} = p$ and $\hat{\sigma} = 1/p$.

i) Time ratio for treatment

```
. lincom _b[treat],eform

 ( 1)  [_t]treat = 0.0

------------------------------------------------------------------------------
      _t |    exp(b)   Std. Err.       z     P>|z|     [95% Conf. Interval]
---------+--------------------------------------------------------------------
     (1) |  1.281358   .1277213     2.49     0.013     1.053964    1.557813
------------------------------------------------------------------------------
```

The estimate of time to return to drug use for the subjects on the longer treatment is 28 percent longer than do subjects on the shorter treatment and it could be as little as 5 percent a much as 56 percent longer with 95 percent confidence.

ii) Time ratio for a 10 year increase in age

```
. lincom 10* _b[age],eform

 ( 1)  10.0 [_t]age = 0.0

------------------------------------------------------------------------------
      _t |    exp(b)   Std. Err.       z     P>|z|     [95% Conf. Interval]
---------+--------------------------------------------------------------------
     (1) |  1.162365   .0936404     1.87     0.062     .9925893    1.36118
------------------------------------------------------------------------------
```

The estimated time ratio is 1.16 with 95% confidence interval between 0.99 and 1.36. The interpretation is that for every 10 year increase in age the time to return to drug use increases by 16 percent and it could be a 1 percent decrease or up to a 36 increase with 95 percent confidence.

> (b) *Using the fitted model compute point and 95 percent confidence interval estimates of the hazard ratio for treatment and for a 10-year change in age. Interpret these estimates within the context of the UIS.*

```
. streg treat age, nolog dist(weibull) nohr

        failure _d:  censor
   analysis time _t:  time


Weibull regression -- log relative-hazard form


No. of subjects =         623                 Number of obs    =         623
No. of failures =         504
Time at risk    =      146816
                                              LR chi2(2)       =        9.63
Log likelihood  =   -1017.0837                Prob > chi2      =      0.0081

------------------------------------------------------------------------------
        _t |      Coef.   Std. Err.      z    P>|z|     [95% Conf. Interval]
-----------+------------------------------------------------------------------
     treat |  -.2219731   .0892721    -2.49   0.013    -.3969431    -.047003
       age |   -.013471   .0072085    -1.87   0.062    -.0275995    .0006575
     _cons |  -4.518468   .3018559   -14.97   0.000    -5.110094   -3.926841
-----------+------------------------------------------------------------------
     /ln_p |  -.1105534   .0370026    -2.99   0.003    -.1830772   -.0380296
-----------+------------------------------------------------------------------
         p |   .8953385   .0331299                      .8327038    .9626844
       1/p |   1.116896   .0413281                      1.038762    1.200907
------------------------------------------------------------------------------
```

We note that the relationship between the coefficients in the above table and those presented earlier for the accelerated failure time parametrization is $\hat{\theta}_1 = \dfrac{-\hat{\beta}_1}{\hat{\sigma}}$ or for treatment $-0.2219731 = \dfrac{-0.2479208}{1.116896}$.

i) Hazard ratio for treatment:

```
. lincom _b[treat],hr

 ( 1)  [_t]treat = 0.0

------------------------------------------------------------------------------
        _t | Haz. Ratio   Std. Err.      z    P>|z|     [95% Conf. Interval]
-----------+------------------------------------------------------------------
       (1) |   .8009369   .0715013    -2.49   0.013     .6723723    .9540845
------------------------------------------------------------------------------
```

Subjects on the longer treatment are returning to drug use at a rate that is estimated to be 20 percent less than subjects on the shorter treatment and the rate could be as little as 5 percent to as much as 33 percent less with 95 percent confidence.

ii) Hazard ratio for a 10 year increase in age

```
. lincom 10*_b[age],hr

 ( 1)  10.0 [_t]age = 0.0

------------------------------------------------------------------------------
          _t |  Haz. Ratio   Std. Err.      z    P>|z|     [95% Conf. Interval]
-------------+----------------------------------------------------------------
         (1) |   .8739695    .0630005    -1.87   0.062     .758817    1.006597
------------------------------------------------------------------------------
```

The estimated hazard ratio is 0.87, and 95% confidence interval is 0.759, 1.007.
The interpretation is that for every 10 year increase in age the rate of returning to drug use decreases by 13 percent and it could be as much as a 24 percent decrease or up to a 0.7 percent increase with 95 percent confidence.

(c) *Compare the time ratio and hazard ratio estimates computed in problems 2(a) and 2(b). In particular, which estimate, time or hazard ratio, do you think would be more easily understood by non-statistically trained study staff?*

The relationship between the coefficients under the two forms of the model is shown above, i.e. $\hat{\theta}_1 = \dfrac{-\hat{\beta}_1}{\hat{\sigma}}$. Which form is easier to interpret depends, as it did in the exponential model, on the statistical background of the investigator and the goals of the analysis. Since time is easier to conceptualize than a rate, the time ratios may be more meaningful to less statistically oriented investigators.

(d) *Why can the fitted Weibull regression model be used to compute both time and hazard ratio estimates?*

The Weibull model is one of only two that can be expressed as both a proportional hazards model and an accelerated failure time model. The other model is the exponential model.

3. *Which model, the exponential regression model fit in problem 1 or the Weibull regression model fit in problem 2, is the better fitting model? Justify your response using plots and statistical tests.*

i) Shape parameter in Weibull model is significant

At the simplest level, the fact that the shape parameter, σ, in the Weibull model is significant ($p = .003$) implies that the Weibull is the preferred model.

ii) Goodness-of- fit test using Cox-Snell residuals

The plots of the Cox-Snell residuals indicate that both the exponential and Weibull regression models seem to fit about the same. Fit is slightly better with the Weibull although neither does very well at the high cumulative hazard levels.

<u>Exponential distribution</u>

To evaluate the fit we use the do file survfit.do used in the previous chapter.

```
. quietly streg treat age, nolog dist(weibull) nohr

. predict Me_t,mgale

  do survfit Me_t 10

Ereg:  likelihood-ratio test                    chi2(9)     =     29.08
                                                Prob > chi2 =    0.0006

. sort group

. by group : list Obs Exp zgroup pzgroup if _n==_N
```

-> group = 1				
	Obs	Exp	zgroup	pzgroup
63.	48	47.43634	.0818386	.9347751

-> group = 2				
	Obs	Exp	zgroup	pzgroup
126.	47	51.94105	-.6855888	.4929725

-> group = 3				
	Obs	Exp	zgroup	pzgroup
189.	47	50.71761	-.5220162	.6016591

-> group = 4				
	Obs	Exp	zgroup	pzgroup
251.	56	42.59275	2.054338	.039943

-> group = 5				
	Obs	Exp	zgroup	pzgroup
314.	48	60.36692	-1.591703	.1114514

-> group = 6				
	Obs	Exp	zgroup	pzgroup
377.	55	40.06886	2.358791	.0183346

-> group = 7				
	Obs	Exp	zgroup	pzgroup
439.	52	48.66969	.4773698	.6330988

-> group = 8				
	Obs	Exp	zgroup	pzgroup
502.	53	52.17598	.1140787	.9091754

-> group = 9				
	Obs	Exp	zgroup	pzgroup
565.	53	52.04434	.1324694	.894613

-> group = 10				
	Obs	Exp	zgroup	pzgroup
628.	49	57.98646	-1.180118	.2379533

150 CHAPTER 8 SOLUTIONS

The Grønnesby-Borgan decile of risk goodness-of-fit test is highly significant, $p = 0.001$. The table of within decile results shows significant departure from fit in deciles 4 and 6. Thus we cannot conclude that the exponential regression model fits.

We show the Arjas plot by quartile of risk below.
Note that there is a departure from fit in the plots for risk groups 2 and 3 providing further evidence of lack-of-fit for the exponential model.

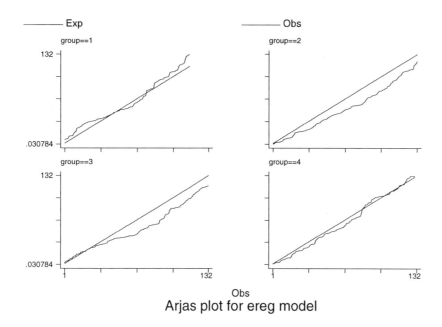

Arjas plot for ereg model

Weibull distribution

```
. quietly streg treat age, nolog dist(weibull) nohr
. predict Mw_t,mgale

. do survfit Mw_t 10
Weibull:  likelihood-ratio test                    chi2(9)      =     25.93
                                                   Prob > chi2  =    0.0021
. sort group
. by group : list Obs Exp zgroup pzgroup if _n==_N
```

-> group = 1

	Obs	Exp	zgroup	pzgroup
63.	48	47.5369	.067167	.9464487

-> group = 2

	Obs	Exp	zgroup	pzgroup
126.	47	51.30663	-.6012434	.5476779

-> group = 3

	Obs	Exp	zgroup	pzgroup
189.	47	50.59746	-.5057452	.6130355

-> group = 4

	Obs	Exp	zgroup	pzgroup
251.	56	43.51306	1.892981	.0583604

-> group = 5

	Obs	Exp	zgroup	pzgroup
314.	48	59.77798	-1.523352	.1276707

-> group = 6

	Obs	Exp	zgroup	pzgroup
377.	55	41.00458	2.185596	.0288452

-> group = 7

	Obs	Exp	zgroup	pzgroup
439.	52	48.68829	.4746141	.6350621

-> group = 8

	Obs	Exp	zgroup	pzgroup
502.	53	52.28345	.0990977	.9210607

-> group = 9

	Obs	Exp	zgroup	pzgroup
565.	53	52.08841	.1263081	.8994881

-> group = 10

	Obs	Exp	zgroup	pzgroup
628.	49	57.20325	-1.084615	.2780924

The Grønnesby-Borgan decile of risk goodness-of-fit test is highly significant, $p = 0.002$. The table of within decile results shows marginally significant departure from fit in decile 4 and a significant departure in decile 6. Thus, we cannot conclude that the Weibull regression model fits.

We show the Arjas plot by quartile of risk below.

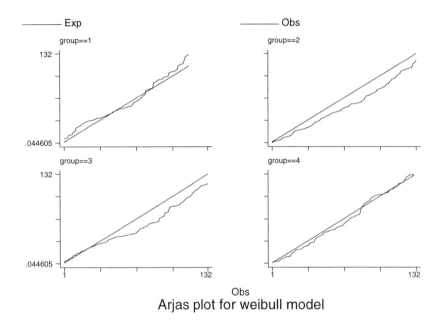

Arjas plot for weibull model

Note that there is a departure from fit in the plots for risk groups 2 and 3 providing further evidence of lack-of-fit for the Weibull model.

Thus we conclude that neither the exponential nor the Weibull models fit the data particularly well.

4. *Verify that the appropriate scale for the covariate AGE in the exponential regression model fit to the data from the HMO-HIV+ study (results shown in Table 8.2) is linear.*

```
. fracpoly ereg TIME AGE DRUG, dead( CENSOR) compare
........
-> gen double IAGE__1 = X^-2-.0769 if e(sample)
-> gen double IAGE__2 = X^3-46.93 if e(sample)
   (where: X = AGE/10)

Iteration 0:   log likelihood = -157.25531
Iteration 1:   log likelihood = -145.24714
Iteration 2:   log likelihood = -129.89625
Iteration 3:   log likelihood = -129.51283
Iteration 4:   log likelihood = -129.51038
Iteration 5:   log likelihood = -129.51038

Exponential regression -- entry time 0
log expected-time form                          Number of obs   =        100
                                                LR chi2(3)      =      55.49
Log likelihood = -129.51038                     Prob > chi2     =     0.0000
------------------------------------------------------------------------------
        TIME |    Coef.   Std. Err.      z    P>|z|    [95% Conf. Interval]
-------------+----------------------------------------------------------------
     IAGE__1 |  9.782858   4.894328     2.00   0.046    .1901509    19.37557
     IAGE__2 |  -.010121   .0063389    -1.60   0.110   -.0225451    .0023031
        DRUG | -1.02158    .2244282    -4.55   0.000   -1.461451   -.5817092
       _cons |  2.789665   .1714077    16.28   0.000    2.453713    3.125618
------------------------------------------------------------------------------
Deviance: 259.021. Best powers of AGE among 44 models fit: -2 3.

Fractional polynomial model comparisons:
------------------------------------------------------------------------------
AGE             df       Deviance    Gain    P(term)  Powers
------------------------------------------------------------------------------
Not in model    0        293.584      --       --
Linear          1        260.794     0.000    0.000   1
m = 1           2        260.155     0.639    0.424   -.5
m = 2           4        259.021     1.773    0.567   -2 3
------------------------------------------------------------------------------
```

Yes. We confirm that the appropriate scale for the covariate AGE in the exponential regression model fit is linear. Gain for $m = 1$ and $m = 2$ model is small compared to linear model.

The Grambsch-Therneau-Fleming smoothed residual plot is shown in the figure below.

- `quietly streg AGE DRUG, dist(exp) nohr`
- `predict Me_t,mgale`
- `gtf_plot CENSOR Me_t AGE`

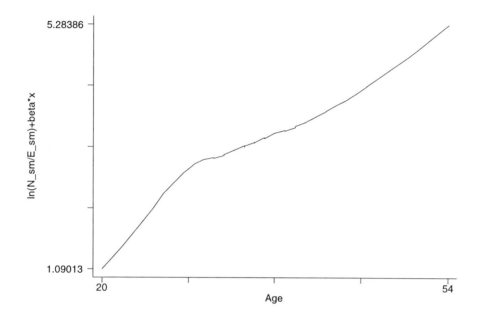

The smoothed plot is quite linear supporting treating age as linear in the model.

5. *Using the results from the fitted Weibull model shown in Table 8.4, compute point and 95 percent confidence limits for the time ratio for history of IV drug use and a 10-year increase in age.*

```
. streg AGE DRUG, dist(weibull) time nolog        Model in Table 8.4
        failure _d:  CENSOR
   analysis time _t:  TIME
Weibull regression -- accelerated failure-time form
No. of subjects =         100                 Number of obs   =        100
No. of failures =          80
Time at risk    =        1136
                                              LR chi2(2)      =      52.05
Log likelihood  =  -128.50229                 Prob > chi2     =     0.0000
------------------------------------------------------------------------------
      _t |      Coef.   Std. Err.      z    P>|z|     [95% Conf. Interval]
---------+--------------------------------------------------------------------
     AGE |  -.0907665    .013616    -6.67   0.000    -.1174534   -.0640796
    DRUG |  -1.049168   .1889778    -5.55   0.000    -1.419558   -.6787786
   _cons |    6.14794   .5107206    12.04   0.000     5.146946    7.148934
---------+--------------------------------------------------------------------
   /ln_p |   .1750802   .0860646     2.03   0.042     .0063967    .3437637
---------+--------------------------------------------------------------------
       p |   1.191342   .1025323                      1.006417    1.410245
     1/p |   .8393897   .0722417                       .7090965    .9936237
------------------------------------------------------------------------------
```

(i) History of drug use

```
. lincom _b[ DRUG], eform

 ( 1)  [_t]DRUG = 0.0

------------------------------------------------------------------------------
      _t |     exp(b)   Std. Err.      z    P>|z|     [95% Conf. Interval]
---------+--------------------------------------------------------------------
     (1) |   .3502289   .0661855    -5.55   0.000     .2418208    .5072362
------------------------------------------------------------------------------
```

Estimates

Point estimate is 0.35 and 95 percent confidence limits for the time ratio for history of IV drug use is (0.24, 0.51).

Interpretation

After controlling for age, the survival times for subjects with a history of IV drug use are estimated to be 35 % of those for subjects without a history of IV drug use and they could as much as 24 percent to 49 percent with 95 percent confidence.

(ii) 10-year increase in age.

```
. lincom 10*_b[ AGE], eform

 ( 1)  10.0 [_t]AGE = 0.0

------------------------------------------------------------------------------
         _t |     exp(b)   Std. Err.      z    P>|z|     [95% Conf. Interval]
------------+-----------------------------------------------------------------
        (1) |   .4034652   .0549359    -6.67   0.000     .3089629    .5268728
------------------------------------------------------------------------------
```

Estimates
The point estimate is 0.4 and 95 percent confidence limits for the time ratio for 10 year increase in age are (0.31, 0.53).

Interpretation
The interpretation is that after controlling for history of IV drug use, the survival times for subjects with every 10 year increase in age the time to death decrease by 40 percent. In other words, the effect of a 10 year increase in age is to reduce survival time by about 60 %. With 95% confidence, the estimate decrease could be between 47% and 69%.

6. *Fit an exponential regression model to the UIS data using the covariates in the main effects proportional hazards model shown in Table 5.5.*

   (a) *Assess the scale of AGE, BECKTOTA and NDRUGTX. Are the results the same or different from those obtained for the proportional hazards model in Chapter 5?*

To assess the scale of continuous covariates, you may use fractional polynomial method and/or residual-based plots. However, it is hard to come up with a parametric function describing the shape of the residual-based plots. Since fractional polynomials suggest the functional form for non-linearly scaled continuous covariates, we will use **fracpoly** procedure here.

AGE

```
. fracpoly ereg  time age becktota ndrugtx ivhx_3 race treat site, dead(censor)
compare
-> gen double Ibeck__1 = becktota-17.37 if e(sample)
-> gen double Indru__1 = ndrugtx-4.543 if e(sample)
........
-> gen double Iage__1 = X^3-33.96 if e(sample)
-> gen double Iage__2 = X^3*ln(X)-39.9 if e(sample)
   (where: X = age/10)
Exponential regression -- entry time 0
log expected-time form                        Number of obs   =        575
                                              LR chi2(8)      =      54.20
Log likelihood = -887.84706                   Prob > chi2     =     0.0000
```

time	Coef.	Std. Err.	z	P>|z|	[95% Conf. Interval]
Iage__1	-.0024981	.032273	-0.08	0.938	-.0657521 .0607559
Iage__2	.0066271	.0200931	0.33	0.742	-.0327547 .0460088
Ibeck__1	-.0080797	.0049503	-1.63	0.103	-.0177822 .0016227
Indru__1	-.032453	.0080023	-4.06	0.000	-.0481371 -.0167688
ivhx_3	-.3017508	.1058696	-2.85	0.004	-.5092514 -.0942501
race	.2388531	.1153036	2.07	0.038	.0128623 .464844
treat	.2280489	.0935293	2.44	0.015	.0447349 .411363
site	.0498977	.108293	0.46	0.645	-.1623528 .2621481
_cons	5.579493	.1059059	52.68	0.000	5.371922 5.787065

Deviance: 1775.694. Best powers of age among 44 models fit: 3 3.
Fractional polynomial model comparisons:

age	df	Deviance	Gain	P(term)	Powers
Not in model	0	1788.944	--	--	
Linear	1	1776.905	0.000	0.001	1
m = 1	2	1775.805	1.100	0.294	3
m = 2	4	1775.694	1.211	0.946	3 3

The p-values reported in the table for the best $m = 1$ and $m = 2$ models are large. Thus, we conclude the best model is linear in age.

BECKTOTA

```
. fracpoly ereg time becktota age ndrugtx ivhx_3 race treat site, dead( censor)
compare

-> gen double Iage__1 = age-32.38 if e(sample)
-> gen double Indru__1 = ndrugtx-4.543 if e(sample)
........
-> gen double Ibeck__1 = X^3-5.284 if e(sample)
-> gen double Ibeck__2 = X^3*ln(X)-2.932 if e(sample)
   (where: X = (becktota+.0499992370605469)/10)

Iteration 0:    log likelihood = -914.94481

Exponential regression -- entry time 0
log expected-time form                          Number of obs   =       575
                                                LR chi2(8)      =     54.90
Log likelihood = -887.49247                     Prob > chi2     =    0.0000

------------------------------------------------------------------------------
        time |      Coef.   Std. Err.      z    P>|z|     [95% Conf. Interval]
-------------+----------------------------------------------------------------
    Ibeck__1 |  -.0382544    .019657    -1.95   0.052    -.0767814    .0002726
    Ibeck__2 |   .025485    .0142856    1.78   0.074    -.0025142    .0534842
     Iage__1 |   .0271062   .0079893    3.39   0.001     .0114475    .042765
    Indru__1 |  -.0316828   .0080631   -3.93   0.000    -.0474862   -.0158794
      ivhx_3 |  -.2925527   .1063188   -2.75   0.006    -.5009337   -.0841717
        race |   .2437154   .1155117    2.11   0.035     .0173167    .4701141
       treat |   .2216744   .0937496    2.36   0.018     .0379285    .4054203
        site |   .0706478   .1081701    0.65   0.514    -.1413616    .2826573
       _cons |   5.611332   .1032886   54.33   0.000     5.40889     5.813774
------------------------------------------------------------------------------
Deviance: 1774.985. Best powers of becktota among 44 models fit: 3 3.

Fractional polynomial model comparisons:
---------------------------------------------------------------
becktota        df    Deviance    Gain    P(term) Powers
---------------------------------------------------------------
Not in model    0     1779.515    --      --
Linear          1     1776.905    0.000   0.106   1
m = 1           2     1776.905    0.000   1.000   1
m = 2           4     1774.985    1.920   0.383   3 3
---------------------------------------------------------------
```

The *p*-values reported in the table for the best $m = 1$ and $m = 2$ models are large. Thus, we conclude the best model is linear in BECKTOTA.

NDRUGTX

```
. fracpoly ereg time ndrugtx becktota age ivhx_3 race treat site, dead( censor)
compare

-> gen double Ibeck__1 = becktota-17.37 if e(sample)
-> gen double Iage__1 = age-32.38 if e(sample)
........
-> gen double Indru__1 = X^-1-1.804 if e(sample)
-> gen double Indru__2 = X^-1*ln(X)+1.065 if e(sample)
   (where: X = (ndrugtx+1)/10)

Exponential regression -- entry time 0
log expected-time form                          Number of obs   =        575
                                                LR chi2(8)      =      61.97
Log likelihood = -883.96227                     Prob > chi2     =     0.0000

------------------------------------------------------------------------------
        time |     Coef.   Std. Err.      z    P>|z|     [95% Conf. Interval]
-------------+----------------------------------------------------------------
     Indru__1|  .5821751   .1235348     4.71   0.000     .3400514    .8242988
     Indru__2|  .2151567   .0479275     4.49   0.000     .1212206    .3090929
     Ibeck__1| -.0087721   .0049831    -1.76   0.078    -.0185388    .0009946
     Iage__1 |  .0300095   .0080816     3.71   0.000     .0141699    .0458492
       ivhx_3| -.2989177   .1076583    -2.78   0.005    -.5099241   -.0879113
         race|  .2555536   .1156363     2.21   0.027     .0289107    .4821966
        treat|  .2093979   .093516      2.24   0.025     .02611      .3926859
         site|  .0787505   .1092936     0.72   0.471    -.135461     .2929621
        _cons|  5.561626   .1081011    51.45   0.000     5.349751    5.7735
------------------------------------------------------------------------------
Deviance: 1767.925. Best powers of ndrugtx among 44 models fit: -1 -1.

Fractional polynomial model comparisons:
------------------------------------------------------------------
ndrugtx       df    Deviance    Gain   P(term)  Powers
------------------------------------------------------------------
Not in model  0     1790.908    --     --
Linear        1     1776.905    0.000  0.000    1
m = 1         2     1775.717    1.188  0.276    .5
m = 2         4     1767.925    8.980  0.020    -1 -1
------------------------------------------------------------------
```

The best power when NDRUGTX enters the model with a single term is 0.5. The approximate partial likelihood ratio test comparing the use of linear term to square root term is $G = 1776.905 - 1775.717 = 1.188$, and the reported p-value is 0.276. Thus, we conclude that a model using the square root of NDRUGTX is not better than a model using NDRUGTX as a linear term.

The best powers when NDRUGTX enters the model with two terms, $m = 2$, is (-1, -1). The partial likelihood ratio test of this model versus the linear model is $G = 1776.905 - 1767.925 = 8.98$ and its significance is $p = \Pr[\chi^2(3) \geq 8.98] = 0.0296$. The $p = 0.02$ in the table is for comparing the $m = 1$ model to the $m = 2$ model. Based on these p-values we conclude $m = 2$ model may be an improvement over the linear model. However before using this transformation we should check the residual plot to make sure it is not being determined by just few data points.

. **fracplot**

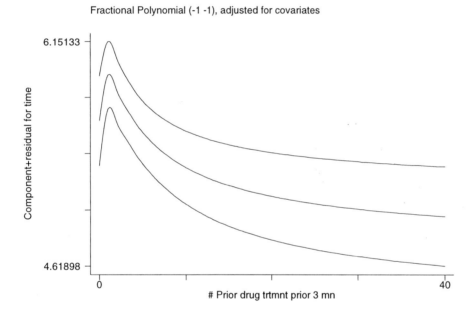

Thus we find that the results of scale selection for age, becktota, and ndrugtx are the same as with the proportional hazards model in Chapter 5.

(b) *Using the correctly scaled model from problem 6(a), assess the need for interactions in the exponential regression model. Are the interactions the same or different from those selected in Chapter 5 for the proportional hazards model?*

Note:
To make the notation simpler we use the same fractional polynomial variables used in Chapters 5 and 6 shown below.

```
. gen x=( ndrugtx+1)/10
(17 missing values generated)
. gen ndrugfp1=1/x
(17 missing values generated)
. gen ndrugfp2= ndrugfp1*ln(x)
(17 missing values generated)
```

Now we fit the main effects model that is used as the comparison model to check for interactions.

```
. stereg age becktota  ndrugfp1 ndrugfp2  ivhx_3 race treat site, time nolog

        failure _d:  censor
   analysis time _t:  time


Exponential regression -- log expected-time form

No. of subjects =         575              Number of obs   =        575
No. of failures =         464
Time at risk    =      138900
                                           LR chi2(8)      =      61.97
Log likelihood  =   -883.96227             Prob > chi2     =     0.0000

------------------------------------------------------------------------------
      _t  |     Coef.    Std. Err.      z     P>|z|    [95% Conf. Interval]
---------+--------------------------------------------------------------------
     age  |   .0300095    .0080816     3.71   0.000     .0141699    .0458492
 becktota |  -.0087721    .0049831    -1.76   0.078    -.0185388    .0009946
 ndrugfp1 |   .5821752    .1235348     4.71   0.000     .3400515    .8242989
 ndrugfp2 |   .2151567    .0479275     4.49   0.000     .1212206    .3090929
   ivhx_3 |  -.2989177    .1076583    -2.78   0.005    -.5099241   -.0879113
     race |   .2555536    .1156363     2.21   0.027     .0289107    .4821966
    treat |   .2093979     .093516     2.24   0.025       .02611    .3926859
     site |   .0787505    .1092936     0.72   0.471     -.135461    .2929621
    _cons |   3.920902    .3640296    10.77   0.000     3.207417    4.634387
------------------------------------------------------------------------------
```

(i) Interactions with AGE

```
. gen agexbeck= age* becktota            Create interaction terms
(37 missing values generated)


. stereg age becktota ndrugfp1 ndrugfp2 ivhx_3 race treat site agexbeck, time nolog
noshow

Exponential regression -- log expected-time form
No. of subjects =         575                   Number of obs   =         575
No. of failures =         464
Time at risk    =      138900
                                                LR chi2(9)      =       61.97
Log likelihood  =   -883.96022                  Prob > chi2     =      0.0000

------------------------------------------------------------------------------
         _t |      Coef.   Std. Err.      z    P>|z|     [95% Conf. Interval]
------------+----------------------------------------------------------------
        age |   .0291455   .0157096     1.86   0.064    -.0016449    .0599358
   becktota |  -.0103753   .0254982    -0.41   0.684    -.0603509    .0396003
   ndrugfp1 |   .5821177   .1235298     4.71   0.000     .3400038    .8242316
   ndrugfp2 |   .2151305   .0479263     4.49   0.000     .1211967    .3090643
     ivhx_3 |  -.2993843   .1078917    -2.77   0.006    -.5108481   -.0879204
       race |   .2560111   .1158589     2.21   0.027     .0289318    .4830904
      treat |   .2091813   .0935778     2.24   0.025     .0257722    .3925904
       site |   .0783416   .1094881     0.72   0.474    -.1362511    .2929343
   agexbeck |   .0000497   .0007748     0.06   0.949     -.001469    .0015683
      _cons |   3.949245    .572662     6.90   0.000     2.826848    5.071642
------------------------------------------------------------------------------
```

The best test to use is the likelihood ratio that compares the above model to the main effects model. However, in this case the p-value for the Wald test for the interaction is so large, $p = 0.949$, that it is nearly certain the two tests would show the interaction is not significant. Thus we will not perform the likelihood ratio test unless the p-value for the Wald test is borderline significant or not significant. No significant interaction between age and becktota. The procedure for all interactions is similar. We present the results for those interactions that are significant rather than show all the output.

AGE by SITE Interaction:

```
. gen agexsite=age*site

. stereg age becktota  ndrugfp1 ndrugfp2  ivhx_3 race treat site agexsite, time nolog
noshow

Exponential regression -- log expected-time form
No. of subjects =          575                    Number of obs   =        575
No. of failures =          464
Time at risk    =       138900
                                                  LR chi2(9)      =      67.33
Log likelihood  =    -881.27858                   Prob > chi2     =     0.0000
------------------------------------------------------------------------------
         _t |      Coef.   Std. Err.       z    P>|z|     [95% Conf. Interval]
------------+----------------------------------------------------------------
        age |   .0426698   .0098268     4.34   0.000     .0234097    .0619299
   becktota |  -.0083024   .0049671    -1.67   0.095    -.0180377    .001433
   ndrugfp1 |   .6166171   .1244569     4.95   0.000     .372686     .8605482
   ndrugfp2 |   .2292366   .0483219     4.74   0.000     .1345274    .3239458
     ivhx_3 |  -.2913655   .1073483    -2.71   0.007    -.5017643   -.0809668
       race |   .2749669   .115861      2.37   0.018     .0478835    .5020503
      treat |   .2366431   .0941501     2.51   0.012     .0521124    .4211739
       site |   1.277209   .5296849     2.41   0.016     .239046     2.315373
   agexsite |  -.0372365   .0159786    -2.33   0.020    -.0685539   -.005919
      _cons |   3.43397    .4192465     8.19   0.000     2.612262    4.255678
------------------------------------------------------------------------------
```

There is a significant interaction between age and site

```
. gen agexfp1= age* ndrugfp1
(18 missing values generated)
. gen agexfp2= age* ndrugfp
(18 missing values generated)


. stereg age becktota  ndrugfp1 ndrugfp2  ivhx_3 race treat site  agexfp1 agexfp2,
time nolog noshow

Exponential regression -- log expected-time form
No. of subjects =         575                    Number of obs   =        575
No. of failures =         464
Time at risk    =      138900
                                                 LR chi2(10)     =      69.44
Log likelihood  =  -880.22282                    Prob > chi2     =     0.0000

------------------------------------------------------------------------------
        _t |      Coef.   Std. Err.      z    P>|z|     [95% Conf. Interval]
-----------+------------------------------------------------------------------
       age |   .0422675   .0274592     1.54   0.124    -.0115515    .0960865
  becktota |  -.0098072   .0049999    -1.96   0.050    -.0196068   -7.51e-06
  ndrugfp1 |   .6014473   .6406526     0.94   0.348    -.6542087    1.857103
  ndrugfp2 |   .1343724    .250418     0.54   0.592    -.3564378    .6251827
    ivhx_3 |  -.2890915   .1070174    -2.70   0.007    -.4988417   -.0793412
      race |   .2827499   .1158811     2.44   0.015     .0556271    .5098727
     treat |   .2060876   .0935804     2.20   0.028     .0226734    .3895017
      site |   .0736331   .1094103     0.67   0.501     -.140807    .2880733
   agexfp1 |   .0010982   .0194339     0.06   0.955    -.0369915     .039188
   agexfp2 |   .0032797   .0076383     0.43   0.668    -.0116912    .0182506
     _cons |   3.455532   .9341875     3.70   0.000     1.624558    5.286505
------------------------------------------------------------------------------


. lrtest, saving(0)

. quietly stereg age becktota  ndrugfp1 ndrugfp2  ivhx_3 race treat site, time

. lrtest
Ereg:  likelihood-ratio test                         chi2(2)    =        7.48
                                                     Prob > chi2 =      0.0238
```

We note that the likelihood ratio test is significant even though the Wald tests are not significant. We performed the LR test in this case as we showed in Chapter 5 that the two fractional polynomial variables ndrugfp1 and ndrugfp2 are highly correlated. Thus we conclude that there may be significant interaction between age and number of prior drug treatments. As was the case in Chapter 5 it is likely best accounted for using a single term like agexfp1.

The next significant interaction is between race and site.

Race by Site Interaction.

```
. gen racexsite = race*site

. stereg age becktota  ndrugfp1 ndrugfp2  ivhx_3 race treat site  racexsite, time
nolog noshow

Exponential regression -- log expected-time form
No. of subjects =         575                   Number of obs   =        575
No. of failures =         464
Time at risk    =      138900
                                                LR chi2(9)      =      75.27
Log likelihood  =   -877.30912                  Prob > chi2     =     0.0000
------------------------------------------------------------------------------
        _t |     Coef.    Std. Err.       z     P>|z|     [95% Conf. Interval]
-----------+------------------------------------------------------------------
       age |   .0322163    .0081405     3.96    0.000     .0162613    .0481713
  becktota |  -.0086583    .0049786    -1.74    0.082    -.0184162    .0010995
  ndrugfp1 |   .6054945    .1234714     4.90    0.000      .363495     .847494
  ndrugfp2 |   .2232527    .0478963     4.66    0.000     .1293778    .3171277
    ivhx_3 |  -.2663444    .1086464    -2.45    0.014    -.4792873   -.0534014
      race |   .4802493       .1345     3.57    0.000     .2166342    .7438644
     treat |   .2240298    .0936035     2.39    0.017     .0405703    .4074894
      site |   .2662729    .1222979     2.18    0.029     .0265734    .5059724
 racexsite |  -.9416567    .2468769    -3.81    0.000    -1.425527   -.4577869
     _cons |   3.727891    .3675153    10.14    0.000     3.007575    4.448208
------------------------------------------------------------------------------
```

The Wald test is significant so we conclude that there is a significant interaction between race and treatment site.

No other interactions are significant.

Summary

Three interaction terms identified as being significant are age and number of previous drug treatments, age and site, and race and site. These interactions were added to the preliminary main effects model. Based on our earlier observations about the correlation between ndrugfp1 and ndrugfp2 we fit the model with only one interaction with age, agexfp1

```
. stereg age becktota  ndrugfp1 ndrugfp2 ivhx_3 race treat site   agexsite agexfp1
racexsite , time nolog noshow

Exponential regression -- log expected-time form

No. of subjects =            575                  Number of obs   =       575
No. of failures =            464
Time at risk    =         138900
                                                  LR chi2(11)     =     85.76
Log likelihood  =     -872.06657                  Prob > chi2     =    0.0000

------------------------------------------------------------------------------
        _t |      Coef.   Std. Err.      z    P>|z|     [95% Conf. Interval]
-----------+------------------------------------------------------------------
       age |   .0628228    .012578     4.99   0.000     .0381703    .0874753
  becktota |  -.0094293   .0049817    -1.89   0.058    -.0191932    .0003347
  ndrugfp1 |    .898105   .1588787     5.65   0.000     .5867085    1.209502
  ndrugfp2 |   .2567603   .0485984     5.28   0.000     .1615092    .3520114
    ivhx_3 |  -.2604103   .1073027    -2.43   0.015    -.4707196   -.0501009
      race |   .5133752   .1345453     3.82   0.000     .2496714    .7770791
     treat |   .2370068   .0942552     2.51   0.012     .0522701    .4217435
      site |    1.14518   .5466139     2.10   0.036     .0738368    2.216524
  agexsite |  -.0276195   .0166225    -1.66   0.097     -.060199      .00496
   agexfp1 |  -.0067056    .002662    -2.52   0.012     -.011923   -.0014882
 racexsite |  -.9208547   .2475186    -3.72   0.000    -1.405982   -.4357272
     _cons |   2.624281   .4925772     5.33   0.000     1.658847    3.589714
------------------------------------------------------------------------------

. lrtest, saving(0)
```

The *p* - value for the Wald tests suggest that the interaction between and age site may not be important in the larger interactions model. So we exclude it and refit the model.

```
. stereg age becktota  ndrugfp1 ndrugfp2 ivhx_3 race treat site  agexfp1   racexsite ,
time nolog  noshow


Exponential regression -- log expected-time form

No. of subjects =         575                  Number of obs   =        575
No. of failures =         464
Time at risk    =      138900
                                               LR chi2(10)     =      83.03
Log likelihood  =  -873.43093                  Prob > chi2     =     0.0000

------------------------------------------------------------------------------
         _t |     Coef.    Std. Err.     z     P>|z|    [95% Conf. Interval]
------------+-----------------------------------------------------------------
        age |   .0559446    .0117742    4.75   0.000    .0328677    .0790216
   becktota |  -.0097732    .0049879   -1.96   0.050   -.0195494    2.93e-06
   ndrugfp1 |   .9010047    .1578322    5.71   0.000    .5916593     1.21035
   ndrugfp2 |   .2492292    .0483236    5.16   0.000    .1545168    .3439417
     ivhx_3 |  -.2653218    .1073799   -2.47   0.013   -.4757825   -.0548611
       race |   .5091872    .1344506    3.79   0.000    .2456689    .7727055
      treat |   .2173647    .0935859    2.32   0.020    .0339397    .4007897
       site |    .265204    .1220094    2.17   0.030    .0260699    .5043381
    agexfp1 |   -.007388    .0026013   -2.84   0.005   -.0124865   -.0022895
   racexsite|  -.9580827     .246669   -3.88   0.000   -1.441545   -.4746204
      _cons |   2.891177    .4638217    6.23   0.000    1.982103     3.80025
------------------------------------------------------------------------------

. lrtest
Ereg: likelihood-ratio test                    chi2(1)         =       2.73
                                               Prob > chi2     =     0.0986
```

The likelihood ratio test for the omitted interaction between age and site is not significant..

In the above model the interaction between age and ndrugfp1 is significant $p = 0.005$.

At this point we go back and check on the interaction between age and site as its lack of significance could also be due to a correlation. To do this we fit the model excluding agexfp1 and including agexsite.

```
. stereg age becktota  ndrugfp1 ndrugfp2 ivhx_3 race treat site  agexsite racexsite ,
time nolog > noshow


Exponential regression -- log expected-time form

No. of subjects =          575              Number of obs   =        575
No. of failures =          464
Time at risk    =       138900
                                            LR chi2(10)     =      79.64
Log likelihood  =   -875.12238              Prob > chi2     =     0.0000

------------------------------------------------------------------------------
         _t |      Coef.   Std. Err.      z    P>|z|     [95% Conf. Interval]
------------+-----------------------------------------------------------------
        age |   .0434578   .0098168     4.43   0.000     .0242173    .0626984
   becktota |  -.008234   .0049639    -1.66   0.097    -.0179631    .0014952
   ndrugfp1 |   .6361911   .1243548     5.12   0.000     .3924602    .8799219
   ndrugfp2 |   .2359093   .0482823     4.89   0.000     .1412777    .3305409
     ivhx_3 |  -.258703   .1083344    -2.39   0.017    -.4710345   -.0463714
       race |   .4918738   .1347227     3.65   0.000     .2278222    .7559254
      treat |   .2463045    .094124     2.62   0.009     .0618247    .4307842
       site |   1.347036   .5319548     2.53   0.011     .3044241    2.389648
    agexsite|  -.033871   .0161043    -2.10   0.035    -.065435    -.0023071
    racexsite| -.9054867   .2473661    -3.66   0.000    -1.390315   -.420658
      _cons |   3.298485   .4200501     7.85   0.000     2.475202    4.121768
------------------------------------------------------------------------------
```

Thus we have the same question we had in Chapter 5. Namely, which interaction with age do we include? At this point we could go forward with either choice; but to keep things a little simpler we use the above model, which is the exponential regression equivalent of the model in Table 5.11.

(c) *Compute the diagnostic statistics to assess the model obtained in problem 6(b). Explore the effect on the estimates of the coefficients of any influential subjects identified through use of the diagnostic statistics.*

Since STATA does not compute all the diagnostic statistics we use a do file written for this purpose. The do file listed below and its use is described at the end of the file.

```
*do file to generate weibull/exponential regression diagnostics
*ref Collett(1994) p188-189 and Hosmer and Lemeshow (1999) p281-282.
*   version 1.0   13Oct97 DWH
* Keep tract of how Stata creates the score residuals.
* They should be in the form "cumhaz-censor" to yield values
* with opposite signs from the log-hazard or PH parametrization
*   e.g. (-1/s)*(c-H) is score residual output by STATA , where s is the shape
parameter
* Also note that matsize needs to be increase from 40 to bigger than _N
* no missing data
version 6
capture program drop scrres
capture drop SV*
capture drop sc*
capture drop DB*
capture drop db*
capture drop M*
capture drop ld*
capture mat drop S
capture mat drop SIG
capture mat drop DB
capture mat drop SVS
capture mat drop coef
program define scrres
 matrix coef=get(_b)
 local b=colsof(coef)
 if "$S_E_cmd" =="weibull" {
      global S_cmd   "wb"
      gen M$S_cmd=-exp(-coef[1,`b'])*res   }
 else {
        global S_cmd "ex"
        gen M$S_cmd =-res
        }
 local i 0
    while   "`1'"  ~=   "" {
         local   i= `i'+1
         gen sc$S_cmd`i'=(`1')*(res)
           local _xvarl=  "`_xvarl'   sc$S_cmd`i'"
           mac shift
       }
dis "`_xvarl' "
mkmat `_xvarl' , mat(S)
 if "$S_E_cmd" =="weibull" {
 mkmat sig, mat(SIG)
 mat S=S,SIG
}
 end
 scrres `*'
 mat V=get(VCE)
mat DB=S*V
svmat DB
capture program drop multsv
program define multsv
 local a = _N
 local b = colsof(coef)
 mat SVS = J(`a',1,0)
 local i =1
   while `i' <= `a' {
        local j =1
          while `j' <=`b' {
           mat SVS[`i',1] = SVS[`i',1]+DB[`i',`j']*S[`i',`j']
           local j= `j'+1
           } /* end j loop*/
     local i=`i'+1
     } /* end i loop*/
  local j = 1
    while `j' <= `b' {
       quietly {gen DB$S_cmd`j'= DB`j'
```

```
                        drop DB`j'
        local j = `j'+1
        }   /* end  j loop */
end
multsv
svmat SVS
gen ld$S_cmd= SVS1
 gen db$S_cmd=SVS1/((1-SVS1)^2)
drop SVS1
exit

*Use of this do file
   1. run program ereg or weibull in log-time form saving the score residuals
      in res and sig (for weibull)
   2. generate one = 1
   3. do weibdiag x1 x2 x3.... one {this is the variable list in Stata order}

output consists of
       sci = scorres for variable xi
             sig is score residual for -ln(sigma) when weibull used
       DBi = delta beta for variable xi
              last DB is DB for -ln(sigma) when weibull used
       db  = Pregibon form for overall belta beta
       ld= likelihood displacement/Cook Distance Stat
       Mi = martingale residuals = censor-exp(z)
             where z=(ln(t)-xb))/sigm
```

Below we list basic summary statistics for the diagnostic statistics.

Variable	Obs	Mean	Std. Dev.	Min	Max
res	575	2.33e-10	.9842047	-.9912649	3.960398
Mex	575	-7.70e-10	.9842047	-3.960398	.9912649
scex1	575	6.28e-09	31.97166	-43.08662	124.5211
scex2	575	-2.11e-08	20.21091	-36.9283	99.53558
scex3	575	6.79e-09	4.628539	-9.751718	27.39432
scex4	575	2.85e-09	9.416768	-63.07776	22.45416
scex5	575	3.17e-10	.6570986	-.9855042	3.960398
scex6	575	-2.70e-10	.4789754	-.9866931	2.869201
scex7	575	1.71e-10	.6520458	-.9912649	3.676775
scex8	575	-2.66e-10	.5254606	-.9855042	3.301407
scex9	575	3.11e-09	17.22753	-43.08662	100.422
scex10	575	-6.48e-11	.2598649	-.9751718	2.869201
scex11	575	7.70e-10	.9842047	-.9912649	3.960398
DBex1	575	-1.30e-12	.0004579	-.0035378	.0023591
DBex2	575	-1.60e-12	.0002241	-.0012151	.0015703
DBex3	575	9.71e-11	.0055441	-.0404873	.0214401
DBex4	575	3.58e-11	.0021684	-.0139348	.0078757
DBex5	575	7.92e-12	.0049403	-.0189959	.0245895
DBex6	575	1.39e-12	.0060649	-.0188841	.0289544
DBex7	575	-7.49e-12	.0042885	-.0187444	.0145888
DBex8	575	-9.14e-11	.0240468	-.1170838	.1429107
DBex9	575	3.69e-12	.000717	-.0036283	.0037897
DBex10	575	1.41e-11	.0112427	-.0320374	.0994023
DBex11	575	-6.68e-11	.0186921	-.0760301	.1225569
ldex	575	.0228461	.042374	7.86e-12	.3880996
dbex	575	.0297739	.080433	7.86e-12	1.036529

We graphed each of the "Dbex*" diagnostics versus "x" for continuous covariates and used box plots for discrete covariates. The same set of points as noted in Chapter 6 were found to be extreme in these plots. The difference is the plots are flipped on the "y-axis" as the fitted model is on the time scale. The plot shown below shows the likelihood displacement statistic versus the martingale residuals.

172 CHAPTER 8 SOLUTIONS

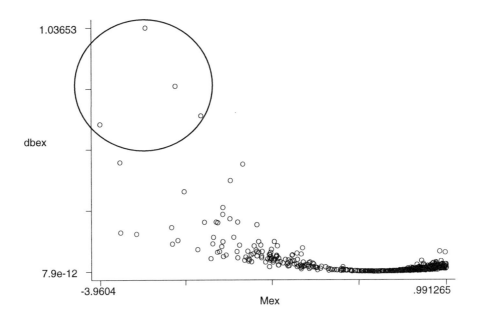

We see four extreme points in the figure. We refit the model excluding these four points and computed the percent change in the coefficients shown below.

age	becktota	ndrugfp1	ndrugfp2	ivhx_3	race
11.929686	22.141208	.94953028	-.18473815	6.5945456	1.9035424

treat	site	agexsite	racexsite	_cons	
26.463353	3.7815491	6.9932953	22.469879	-5.8462636	

We see that the change in the coefficient for becktota, treat and racexsite have changed by more than 20 percent. What is different in this model from the PH model in the text is that the coefficient for treatment has changed and is actually much larger (0.31 vs. 0.25) and remains significant with $p = 0.001$. At this point we should examine the data for these four subjects with the study team to assess possible explanations for the effect of these subjects.

(d)   Assess the overall fit of the model from problem 6(c) via the Grønnesby – Borgan test.

We perform this analysis using the survfit.do file used in Chapter 6. Results are given below

```
. lrtest
Ereg:   likelihood-ratio test                          chi2(9)    =      12.64
                                                       Prob > chi2 =     0.1794
. sort group
. by group : list Obs Exp zgroup pzgroup if _n==_N
```

-> group = 1

	Obs	Exp	zgroup	pzgroup
58.	54	62.12305	-1.030607	.3027253

-> group = 2

	Obs	Exp	zgroup	pzgroup
115.	53	52.47108	.0730184	.9417915

-> group = 3

	Obs	Exp	zgroup	pzgroup
173.	56	44.18185	1.777983	.0754066

-> group = 4

	Obs	Exp	zgroup	pzgroup
230.	48	49.65752	-.2352162	.8140409

-> group = 5

	Obs	Exp	zgroup	pzgroup
288.	51	40.49324	1.651116	.0987148

-> group = 6

	Obs	Exp	zgroup	pzgroup
345.	46	50.41729	-.622108	.5338709

-> group = 7

	Obs	Exp	zgroup	pzgroup
402.	39	50.2821	-1.591047	.1115989

-> group = 8

	Obs	Exp	zgroup	pzgroup
460.	40	44.03241	-.6076854	.5433962

-> group = 9

	Obs	Exp	zgroup	pzgroup
517.	41	38.09042	.471435	.6373301

-> group = 10

	Obs	Exp	zgroup	pzgroup
575.	36	32.25104	.6601442	.5091614

The p-value for the Grønnesby – Borgan test is $p = 0.1794$. Thus we can not reject that the model fits. The table of within decile of risk z-scores and p-values indicates that the observed and expected numbers of events are not significantly different.

The Arjas plot by quartile of risk is shown below and indicates good agreement between the

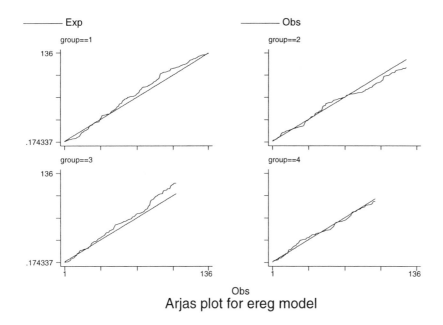

Arjas plot for ereg model

cumulative within group observed and expected number of events

>   (e)   *Assess the adherence of the fitted model to the exponential errors assumption via the plot of the Cox–Snell residuals.*

We will program the steps presented on page 286 of the text.

The estimator of the Cox-Snell residuals from an exponential regression model is obtained by exponentiating the additive residuals on the log-time scale.

```
. predict cs, csn    * equivalently   . gen cs= time*exp(-xb)
            (53 missing values generated)
. drop if cs==.
(53 observations deleted)

. stset   cs, failure(censor)

    failure event:  censor ~= 0 & censor ~= .
obs. time interval:  (0, cs]
 exit on or before:  failure

------------------------------------------------------------------------
      575  total obs.
        0  exclusions
------------------------------------------------------------------------
      575  obs. remaining, representing
      464  failures in single record/single failure data
      464  total analysis time at risk, at risk from t =         0
                              earliest observed entry t =         0
                                   last observed exit t =  3.960398


. sts gen km=s

. gen double H=-ln(km)     Cumulative expected
```

```
. graph H cs cs, c(ll) s(o.) xlab ylab sort
```

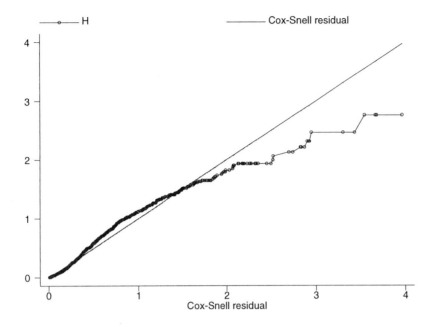

The plot of the Cox-Snell residuals indicates that the exponential regression model seems to fit except at the highest values of the cumulative hazard. He model seems to over estimate the cumulative hazard.

(f)   Describe the estimates of effect of the covariates in the final exponential regression model using time ratios, with 95 percent confidence intervals.

```
. stereg age becktota  ndrugfp1 ndrugfp2 ivhx_3 race treat site agexsite racexsit, tr
noshow nolog

Exponential regression -- log expected-time form

No. of subjects =        575                  Number of obs   =       575
No. of failures =        464
Time at risk    =     138900
                                              LR chi2(10)     =     79.64
Log likelihood  =  -875.12238                 Prob > chi2     =    0.0000
```

_t	Tm. Ratio	Std. Err.	z	P>|z|	[95% Conf. Interval]
age	1.044416	.0102528	4.43	0.000	1.024513 1.064706
becktota	.9917998	.0049232	-1.66	0.097	.9821973 1.001496
ndrugfp1	1.889271	.2349399	5.12	0.000	1.480619 2.410712
ndrugfp2	1.266059	.0611283	4.89	0.000	1.151744 1.391721
ivhx_3	.7720523	.0836398	-2.39	0.017	.624356 .9546873
race	1.635378	.2203225	3.65	0.000	1.255862 2.129581
treat	1.279289	.1204118	2.62	0.009	1.063776 1.538463
site	3.84601	2.045903	2.53	0.011	1.355844 10.90966
agexsite	.9666962	.015568	-2.10	0.035	.93666 .9976955
racexsite	.4043451	.1000213	-3.66	0.000	.2489968 .6566147

i) Time ratio for "ivhx_3" can be read off the above table

The time return to drug use for subject with history of recent IV drug use is estimated to be 23% shorter than do subject with no history of recent IV drug use and it could be as little as 4% to as much as 38% with 95 percent confidence.

ii) Time ratio for "treat" can be read off the above table

The time return to drug use for subjects on the longer treatment is estimated to be 27% longer than for subjects on the shorter treatment and it could be as little as 6% or as much as 54% with 95% confidence.

iii) Time ratio for AGE

Due to the age by site interaction we use **lincom** and compute the TR for a 10 year change within each site.

Site A

```
. lincom _b[ age]*10, eform

 ( 1)  10.0 [_t]age = 0.0

------------------------------------------------------------------------------
          _t | Haz. Ratio   Std. Err.      z    P>|z|     [95% Conf. Interval]
-------------+----------------------------------------------------------------
         (1) |   1.544312    .1516016     4.43   0.000     1.274015    1.871955
------------------------------------------------------------------------------
```

The estimated time ratio is 1.54, and its 95% confidence limits are 1.27 and 1.87 in treatment site A. The time to return to drug use is estimated to be 54% longer for every 10 year increase in age.

Site B

```
. lincom _b[ age]*10+ _b[ agexsite]*10, eform

 ( 1)  10.0 [_t]age + 10.0 [_t]agexsite = 0.0

------------------------------------------------------------------------------
          _t | Haz. Ratio   Std. Err.      z    P>|z|     [95% Conf. Interval]
-------------+----------------------------------------------------------------
         (1) |   1.100614    .1469585     0.72   0.473     .8471863    1.429851
------------------------------------------------------------------------------
```

The estimated time ratio is 1.1 and its 95% confidence limits are 0.85 and 1.43 in treatment site B. Increasing age by 10 years has no significant effect on the time to return to drug use in Site B.

iv) BECKTOTA

We estimate the time ratios for a 10 point increase in BECKTOTA via **lincom**.

```
. lincom 10*_b[ becktota], eform

 ( 1)  10.0 [_t]becktota = 0.0

------------------------------------------------------------------------------
          _t | Haz. Ratio   Std. Err.      z    P>|z|     [95% Conf. Interval]
-------------+----------------------------------------------------------------
         (1) |   .9209591    .0457159    -1.66   0.097     .8355784    1.015064
------------------------------------------------------------------------------
```

The estimated time ratio is 0.92, with 0.05 < p < 0.10. Increasing Beck Score by 10 points produces a marginally significant 8% reduction in the time to return to drug use.

v) The estimates of time ratio for RACE must be done within site due to the interaction.

Site A

The time ratio for RACE at SITE A can be read off the above table

The estimated time ratio is 1.63 and its 95% confidence limits are 1.26 and 2.13 in treatment site B. The estimated time to return to drug use for non-white race subjects is 63% longer than for subjects of white race.

Site B

```
. lincom _b[ race] + _b[ racexsit], efrom

 ( 1)  [_t]race + [_t]racexsite = 0.0

------------------------------------------------------------------------
       _t |  Haz. Ratio   Std. Err.      z    P>|z|   [95% Conf. Interval]
----------+-------------------------------------------------------------
      (1) |   .6612569    .1387397    -1.97   0.049    .4383064   .9976142
------------------------------------------------------------------------
```

The estimated time ratio is 0.66 and its 95% confidence limits are 0.44 and 0.98 in treatment site B. Here we estimate that the time to return to drug use for subjects of non-white race is 34% shorter than subjects of white race.

vi) Time ratio for NDRUGTX

Since NDRUGTX is modeled with two fractional polynomials we cannot use one time ratio to compare an increase in this covariate. Recall that in Chapter 6 we used a graph to estimate the hazard ratio for a one treatment increase and a table to compare various treatment numbers to the minimum of one treatment. In the AFT formulation the maximum occurs at one treatment. Thus we choose here to estimate time ratios for 0, 2, 5 and 10 previous treatments to one treatment. These calculations lend themselves nicely to a spreadsheet. However we need the covariance matrix from the fitted model. This is shown below by using the **vce** command following the fit of the model.

```
. quietly stereg age becktota ndrugfp1 ndrugfp2 ivhx_3 race treat site agexsite
racexsit,time

. vce
           |       age  becktota  ndrugt_1  ndrugt_2     ivhx_3       race      treat
-----------+----------------------------------------------------------------------------
       age |   .000096
  becktota |   2.7e-06   .000025
  ndrugt_1 |   .000259  -.000031   .015464
  ndrugt_2 |   .000092  -.000013   .005944    .002331
    ivhx_3 |  -.000172  -.000057   .001737    .000503    .011736
      race |  -8.4e-06  -.000035   .000546    .00026     .003216    .01815
     treat |   .000078   .000013   .000028   -9.9e-06    .000723    .000013   .008859
      site |   .002821   .000129   .008725    .003628    .005188    .008017   .006441
  agexsite |  -.000089  -3.5e-06  -.000232   -.000096   -.00006    -.000089  -.000177
  racexsit |  -.000064  -2.2e-06  -.001572   -.000536   -.002551   -.01766   -.000811
     _cons |  -.003496  -.00046   -.030461   -.011148   -.003975   -.00564   -.007704

           |      site  agexsite  racexsit      _cons
-----------+-----------------------------------------
      site |   .282976                .
  agexsite |  -.008342   .000259
  racexsit |  -.005729  -.000264   .06119
     _cons |  -.115225   .003362   .010525   .176442
```

The following spreadsheet calculates time ratios for 0, 2, 5, and 10 previous drug treatments vs 1 previous drug treatment. In the spread sheet the column labeled "a" contains the values of the difference in the first factional polynomial at 0, 2, 5 and 10 and the value at 1. The column labeled "b" are the differences in second fractional polynomial. The formulae are shown on pages 323 – 234 in Chapter 6 for the PH model. The equations are the same in this setting. The key equation is (6.29).

NDRUGTX	10/(x+1)	a	b	lincom	TR	se(lincom)
0	10.0000	5.0000	-14.9787	-0.3527	0.7028	0.1388
2	3.3333	-1.6667	4.0339	-0.1087	0.8970	0.0310
5	1.6667	-3.3333	7.1958	-0.4231	0.6550	0.0859
10	0.9091	-4.0909	8.1338	-0.6838	0.5047	0.1321
		lincom lb	lincom ub	tr lb	tr ub	
		-0.6246	-0.0807	0.5355	0.9225	
		-0.1694	-0.0479	0.8441	0.9532	
		-0.5914	-0.2548	0.5536	0.7751	
		-0.9426	-0.4249	0.3896	0.6539	
		coeffs	var/cov			
		0.6362	0.0155	0.0059		
		0.2359		0.0023		

The results are summarized in the table below. We note that the estimated time ratios are approximately the inverse of the hazard ratios in Table 6.8, page 235. This is not unexpected as the exponential regression model is also a PH model. In general we see that subjects with 1,2,5 or 10 previous drug treatments are returning to dug use from 10 to 50 percent sooner than subjects with one previous treatment.

	0	2	5	10
TR	0.70	0.90	0.66	0.50
95% CIE	(0.53, 0.92)	(0.84, 0.05)	(0.55, 0.78)	(0.39, 0.65)

7. *Repeat problem 6 using the Weibull regression model. Is the fit of the Weibull regression superior to the exponential model?*

We do not show the work for this problem as all that is required is to repeat the analysis in problem 6 with the Weibull distribution. Note that when using the do file one must save the scores for the shape parameter in a variable name.

We do show the fit of the basic model.

```
. streg age becktota    ndrugfp1 ndrugfp2 ivhx_3 race treat site agexsite racexsit,
nolog noshow time dist(weibull) score(res,s)
```

Weibull regression -- accelerated failure-time form

No. of subjects =   575   Number of obs   =   575
No. of failures =   464
Time at risk   =   138900
                                                  LR chi2(10)   =   77.09
Log likelihood =   -875.12052   Prob > chi2   =   0.0000

_t	Coef.	Std. Err.	z	P>|z|	[95% Conf. Interval]
age	.0435056	.0098706	4.41	0.000	.0241596 .0628516
becktota	-.0082464	.0049796	-1.66	0.098	-.0180062 .0015134
ndrugfp1	.6367754	.1250115	5.09	0.000	.3917574 .8817934
ndrugfp2	.2361305	.0485301	4.87	0.000	.1410132 .3312477
ivhx_3	-.2589951	.108689	-2.38	0.017	-.4720216 -.0459685
race	.4924565	.1353682	3.64	0.000	.2271398 .7577732
treat	.2465655	.094438	2.61	0.009	.0614704 .4316606
site	1.348693	.5338753	2.53	0.012	.3023164 2.395069
agexsite	-.0339136	.0161568	-2.10	0.036	-.0655804 -.0022469
racexsite	-.9065252	.2485186	-3.65	0.000	-1.393613 -.4194377
_cons	3.295919	.4230897	7.79	0.000	2.466678 4.125159
/ln_p	-.002317	.0379921	-0.06	0.951	-.0767801 .072146
p	.9976856	.0379041			.9260935 1.074812
1/p	1.00232	.0380802			.930395 1.079805

The Wald test for the significance of the scale parameter ($p = .951$) suggests that the Weibull offers no improvement in fit over the exponential regression model examined in problem 6.

To examine the fit we use survfit.do whose output is shown below.

```
.lrtest
Weibull:  likelihood-ratio test                        chi2(9)     =     12.72
                                                       Prob > chi2 =    0.1757

. sort group

. by group : list Obs Exp zgroup pzgroup if _n==_N
```

-> group = 1
	Obs	Exp	zgroup	pzgroup
58.	54	62.1098	-1.029035	.3034632

-> group = 2
	Obs	Exp	zgroup	pzgroup
115.	53	52.47203	.0728861	.9418967

-> group = 3
	Obs	Exp	zgroup	pzgroup
173.	56	44.20146	1.774639	.0759575

-> group = 4
	Obs	Exp	zgroup	pzgroup
230.	48	49.65268	-.2345397	.8145661

-> group = 5
	Obs	Exp	zgroup	pzgroup
288.	51	40.50554	1.648932	.0991616

-> group = 6
	Obs	Exp	zgroup	pzgroup
345.	46	50.41086	-.6212429	.5344398

-> group = 7
	Obs	Exp	zgroup	pzgroup
402.	39	50.27002	-1.589535	.1119396

-> group = 8
	Obs	Exp	zgroup	pzgroup
460.	40	44.02514	-.6066387	.5440907

-> group = 9
	Obs	Exp	zgroup	pzgroup
517.	41	38.09178	.4712073	.6374927

-> group = 10
	Obs	Exp	zgroup	pzgroup
575.	36	32.26069	.6583462	.5103157

The overall test and within decile of risk *p*-values all support model fit.

To examine this further, we examine the plot of the Cox-Snell residuals performed in part 6(e) for the exponential regression model

```
. predict cs, csnell

. stset cs, failure(censor)

. sts gen km=s
. gen double H=-ln(km)
  (53 missing values generated)

. graph H cs cs, c(ll) s(o.) xlab ylab sort
```

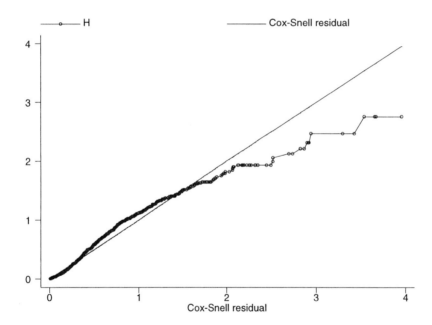

The above figure demonstrates that the Weibull offers no improvement over the exponential regression model. Both models seem to fit adequately but do not show good adherence to distribution assumptions in the Cox-Snell residual plot. Also the fact that the shape parameter in the Weibull fit is not significant suggests that the simpler exponential regression model is just as good.

8. *Fit a log-logistic regression model containing the covariates in the final Weibull regression model from problem 6. Repeat problem 6(f) for the log-logistic model. Is the fit of the log-logistic regression superior to the Weibull or exponential regression models?*

```
. streg age becktota    ndrugfp1 ndrugfp2 ivhx_3 race treat site agexsite racexsit,
dist(loglogistic) time nolog noshow

Log-logistic regression -- accelerated failure-time form

No. of subjects =          575               Number of obs   =         575
No. of failures =          464
Time at risk    =  464.0000008
                                             LR chi2(10)     =        2.50
Log likelihood  =   -851.42545               Prob > chi2     =      0.9909

------------------------------------------------------------------------------
         _t |      Coef.   Std. Err.      z    P>|z|     [95% Conf. Interval]
-------------+----------------------------------------------------------------
        age |   .0013553   .0105187     0.13   0.897    -.019261    .0219716
    becktota |  -.0040505   .0053849    -0.75   0.452    -.0146048    .0065038
    ndrugfp1 |   -.091449   .1371391    -0.67   0.505    -.3602367    .1773387
    ndrugfp2 |  -.0288254   .0530048    -0.54   0.587    -.1327128     .075062
      ivhx_3 |    .005029   .1156593     0.04   0.965     -.221659    .2317171
        race |  -.0022311    .137747    -0.02   0.987    -.2722103    .2677482
       treat |   .0808524   .1003889     0.81   0.421    -.1159063    .2776111
        site |   .2301208   .5781103     0.40   0.691    -.9029546    1.363196
    agexsite |  -.0065979   .0176209    -0.37   0.708    -.0411342    .0279384
    racexsite |   .0057595   .2608374     0.02   0.982    -.5054723    .5169913
       _cons |  -.3492549   .4433341    -0.79   0.431    -1.218174     .519664
-------------+----------------------------------------------------------------
     /ln_gam |  -.3785701   .0390841    -9.69   0.000    -.4551736   -.3019666
-------------+----------------------------------------------------------------
       gamma |     .68484   .0267664                      .6343378    .7393628
------------------------------------------------------------------------------
```

The coefficients are similar to what was found for the exponential and Weibull models. The results of running survfit.do are shown below. The results support overall model fit. There is a significant departure from fit in the 5th decile of risk (p=.009), however the fit in all other deciles is quite good.

```
. lrtest
Llogistic:    likelihood-ratio test                      chi2(9)     =    13.16
                                                         Prob > chi2 =   0.1557

. sort group

. by group : list Obs Exp zgroup pzgroup if _n==_N
```

-> group = 1				
	Obs	Exp	zgroup	pzgroup
58.	45	56.06477	-1.477738	.139478

-> group = 2				
	Obs	Exp	zgroup	pzgroup
115.	43	44.64934	-.2468329	.8050376

-> group = 3				
	Obs	Exp	zgroup	pzgroup
173.	47	54.93922	-1.071116	.2841172

-> group = 4				
	Obs	Exp	zgroup	pzgroup
230.	41	50.89757	-1.387331	.1653408

-> group = 5				
	Obs	Exp	zgroup	pzgroup
288.	52	36.31897	2.602002	.0092681

-> group = 6				
	Obs	Exp	zgroup	pzgroup
345.	47	42.10954	.7536324	.45107

-> group = 7				
	Obs	Exp	zgroup	pzgroup
402.	47	46.30248	.1025069	.9183544

-> group = 8				
	Obs	Exp	zgroup	pzgroup
460.	46	41.72483	.661844	.5080712

-> group = 9				
	Obs	Exp	zgroup	pzgroup
517.	49	44.37022	.6950482	.4870251

-> group = 10				
	Obs	Exp	zgroup	pzgroup
575.	47	53.06924	-.8331308	.404771

The Cox-Snell analysis follows.

```
. predict cs, csn

. stset   cs, failure(censor)

. sts gen km=s

. gen double H=-ln(km)
  (53 missing values generated)

. graph H cs cs, c(ll) s(o.) xlab ylab sort
```

The figure shows that the log-logistic model does offer some improvement in fit over the exponential and Weibull models.

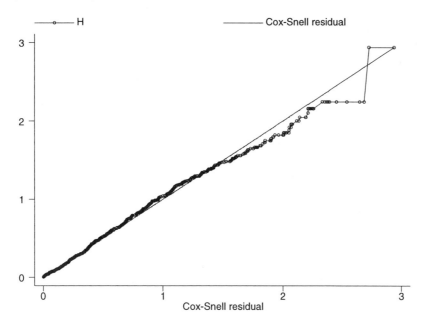

Another way to compare the models is via the AIC criterion shown below.

Model	log likelihood	AIC
Exponential	-875.122	1772.245
Weibull	-875.121	1774.241
log-logistic	-852.501	1729.001

The AIC for the exponential and Weibull are quite similar and are much large than that for the log-logistic model. Thus of the three AFT models the log logistic is the best fitting.

Chapter Nine – Solutions

1. Consider the following hypothetical study of two treatment modalities to reduce the occurrence of muscle soreness among middle-aged men beginning weight training. Study participants were 400 middle-aged men who joined a health club for the specific purpose of weight training. Subjects were randomized into one of two instructional programs designed to prevent muscle soreness. The control treatment consisted of the standard written brochures and instructions used by the health club to explain proper technique, including suggestions for frequency and duration of training. The new method included 1 hour with a personal trainer as well as the brochures. Subjects were followed and the dates on which muscle soreness limited the prescribed workout were recorded. The dates were converted into the number of days between soreness episodes.

The data may be found on the statistical data set web site at the University of Massachusetts, Amherst and the John Wiley web site discussed in Section 1.3 and the Preface. The data are in a file called RECUR.DAT. The variables are: ID (1–400), AGE (years), TREAT (0 = NEW, 1 = CONTROL), TIME0 (day of the previous episode or 0), TIME1 (day of new episode), CENSOR (1 = muscle soreness episode occurred at TIME1, 0 = subject left the study or the study ended at TIME1) and EVENT (1 – 4 muscle soreness episode). The maximum number of episodes observed is 4.

Every study subject had one episode, 386 had two, 324 had three, and 186 had four. Thus the data file has 1296 records and is in the form shown for the counting process recurrent event model shown in Table 9.1. The data for this hypothetical study were generated to have sufficient power to detect particular differences in the models. A careful analysis will uncover these differences.

(i) Codebook

Data file recur.dat has n = 400 subjects. Up to four recurrent events are possible. Data are set out in counting process format where each event is described by the time the previous event took place and the time of the next event or censoring took place.
Variable Name   Codes / Units
ID   Subject Identification   1 - 400
AGE   Age years
TREAT   Treatment Assignment   0 = New 1 = Old
TIME0   Time of Last Episode Days
TIME1   Time of Current Episode Days or censoring
CENSOR   Indicator for Soreness   1 = Episode Number
Episode or Censoring   0 = Censored 1 = Event
EVENT   Soreness Episode Number 1 to at most 4

(ii) The basic set up of the data is listed in the table below.

ID	AGE	TREAT	TIME0	TIME1	CENSOR	EVENT
1	43	0	0	6	1	1
1	43	0	6	9	1	2
1	43	0	9	56	1	3
1	43	0	56	88	1	4
2	43	0	0	42	1	1
2	43	0	42	87	1	2
2	43	0	87	91	0	3
3	41	0	0	15	1	1
3	41	0	15	17	1	2
3	41	0	17	36	1	3
3	41	0	36	112	0	4
:	:	:	:	:	:	:
:	:	:	:	:	:	:
:	:	:	:	:	:	:
400	38	1	0	20	1	1
400	38	1	20	78	1	2
400	38	1	78	80	0	3

To declare data to be survival time data, it is important to understand how variables are entered.

Consider the first patient who had 4 events. This subject had an episode of muscle soreness at 6, 9, 56, and 88 days of follow-up. The first visit was made at time = 6 and this person experience episode of muscle soreness. Since "time0" is the time of last episode days and "time1" is the time of current episode days or censoring, time0 is 0 and time1 is 6 for this visit. The second episode of muscle soreness occurred 9 days. Fort his episode time0 = 6 and time1 = 9.

(iii) Declare data to be survival time data.

How one declares data to be survival-time data depends on the recurrent event model one chooses. We illustrate this in the following four problems.

> (a) *Fit the counting process recurrent event model to these data obtaining both the information matrix-based estimates and the robust estimates of the standard errors of the estimated coefficients for AGE and TREAT.*

In the counting process model, the recurrent events within a subject are assumed to be independent, and a subject contributes to the risk set for an event as long as the subject is under observation at the time each event occurs. For example, the first subject will be in the risk set for any event occurring between 0 and 88 days. The data for the first subject could be thought of as data for four different subjects. The first "subject" has follow-up from 0 to 6 days when his / her event occurs. The second "subject" has delayed entry at 6 days and is followed until 9 days when muscle soreness occurs. The third "subject" has delayed entry at 9 days and is followed until 56 days when the muscle soreness occurs. The fourth "subjects" has delayed entry at 56 and is followed until 88 days when his /her episode of muscle soreness occurs.

CHAPTER 9 SOLUTIONS

The STATA commands for this type of data are shown in the following **stset** command. Note that there are 1296 observations.

```
. stset    time1, failure(censor) time0(time0)

     failure event:  censor ~= 0 & censor ~= .
obs. time interval:  (time0, time1]
 exit on or before:  failure

------------------------------------------------------------------------
      1296  total obs.
         0  exclusions
------------------------------------------------------------------------
      1296  obs. remaining, representing
       939  failures in single record/single failure data
     39904  total analysis time at risk, at risk from t =         0
                                    earliest observed entry t =   0
                                         last observed exit t =   380
```

(i) Information matrix-based estimate of the standard error

```
. stcox age treat, nohr nolog    Information matrix-based est. of the std. error

failure _d:  censor
   analysis time _t:  time1

Cox regression -- Breslow method for ties

No. of subjects =         1296                    Number of obs   =      1296
No. of failures =          939
Time at risk    =        39904
                                                  LR chi2(2)      =     28.99
Log likelihood  =    -5185.4266                   Prob > chi2     =    0.0000
------------------------------------------------------------------------------
     _t |
     _d |      Coef.    Std. Err.      z     P>|z|     [95% Conf. Interval]
--------+---------------------------------------------------------------------
    age |    .0437907    .0108745    4.03   0.000     .0224771    .0651043
  treat |    .2408107    .0658153    3.66   0.000     .1118151    .3698063
------------------------------------------------------------------------------
```

(ii) Robust estimator of the covariance matrix.

The key to the robust calculation is using the efficient score residuals for each subject in the data for the variance calculation.

The **cluster** option implies that one wants to use the robust estimator and specifies a variable on which the clustering is to be based.

```
. stcox age treat, nohr nolog  cluster(id)

        failure _d:  censor
   analysis time _t:  time1

Cox regression -- Breslow method for ties
No. of subjects =       1296                   Number of obs    =     1296
No. of failures =        939
Time at risk    =      39904
                                               Wald chi2(2)     =    53.80
Log likelihood  =   -5185.4266                 Prob > chi2      =   0.0000
                              (standard errors adjusted for clustering on id)

------------------------------------------------------------------------------
         _t |              Robust
         _d |    Coef.   Std. Err.      z    P>|z|     [95% Conf. Interval]
------------+-----------------------------------------------------------------
        age |  .0437907   .0074679    5.86   0.000     .0291539    .0584274
      treat |  .2408107   .0476084    5.06   0.000     .1475       .3341214
------------------------------------------------------------------------------
```

Note that the standard errors calculated by robust method are both smaller, but similar in magnitude to the information matrix based estimator.

```
. lincom _b[treat],hr

 ( 1)  treat = 0.0

------------------------------------------------------------------------------
         _t | Haz. Ratio   Std. Err.      z    P>|z|     [95% Conf. Interval]
------------+-----------------------------------------------------------------
        (1) |   1.27228    .0605712    5.06   0.000     1.158933    1.396713
------------------------------------------------------------------------------
```

(b) Repeat problem 1(a) fitting the conditional A recurrent event model. Does the effect of the covariates depend on the episode number?

Under conditional model A, we assume a subject is not at risk for a subsequent muscle soreness event until a prior episode has occurred. For example, subject 1 is assumed not to be at risk for the second episode until the first episode has occurred, after 6 days of follow-up. Similarly, this subject is assumed not to be at risk for his/her third episode until the second episode has occurred. A stratum variable denoting the particular soreness episode number (EVENT) is used to keep track of the particular event number.

The difference between Conditional model A and Conditional model B is the time scale used. Conditional model A uses time defined by the beginning of the study while the Conditional model B uses time since the previous episode.

We begin by setting and then fitting Conditional model A. Note that the **stset** command is the same as used for the counting process model. The difference is in the use of EVENT as a stratification variable.

```
. stset    time1, failure(censor) time0(time0)
     failure event:   censor ~= 0 & censor ~= .
obs. time interval:   (time0, time1]
 exit on or before:   failure
------------------------------------------------------------------------
      1296  total obs.
         0  exclusions
------------------------------------------------------------------------
      1296  obs. remaining, representing
       939  failures in single record/single failure data
     39904  total analysis time at risk, at risk from t =         0
                                  earliest observed entry t =         0
                                       last observed exit t =       380
```

(i) The fit of Conditional model A

Information matrix based estimates

```
. stcox   age treat, nohr nolog strata( event)
         failure _d:  censor
   analysis time _t:  time1
Stratified Cox regr. -- Breslow method for ties
No. of subjects =         1296                 Number of obs   =       1296
No. of failures =          939
Time at risk    =        39904
                                               LR chi2(2)      =      69.06
Log likelihood  =    -4166.8323                Prob > chi2     =     0.0000
------------------------------------------------------------------------
       _t |
       _d |     Coef.    Std. Err.      z      P>|z|     [95% Conf. Interval]
----------+-------------------------------------------------------------
      age |   .057643    .0109627    5.26    0.000    .0361565    .0791295
    treat |   .472991    .070094     6.75    0.000    .3356092    .6103727
------------------------------------------------------------------------
                                                       Stratified by event
```

Robust estimates

```
. stcox   age treat, nohr nolog cluster(id) strata( event) Robust estimates

        failure _d:  censor
   analysis time _t:  time1
Stratified Cox regr. -- Breslow method for ties
No. of subjects =         1296                Number of obs   =       1296
No. of failures =          939
Time at risk    =        39904
                                              Wald chi2(2)    =      67.40
Log likelihood  =   -4166.8323                Prob > chi2     =     0.0000

                        (standard errors adjusted for clustering on id)
------------------------------------------------------------------------------
          _t |               Robust
          _d |      Coef.   Std. Err.      z    P>|z|     [95% Conf. Interval]
-------------+----------------------------------------------------------------
         age |   .057643    .0102669     5.61   0.000     .0375203    .0777658
       treat |   .472991    .0709855     6.66   0.000     .3338619     .61212
------------------------------------------------------------------------------
                                                            Stratified by event
```

Note that in this model the robust estimates are nearly identical to the information matrix based estimates.

(ii) Test if effect of treatment and age depend on the number of episode.

We assess if the event number modifies the effect of the covariates through inclusion of interaction terms between covariates and the EVENT variable. To be a bit conservative we fit the model using the robust estimator.

```
. xi:stcox i.event|age i.event|treat, nohr nolog strata( event) cluster(id)
i.event          _Ievent_1-4      (naturally coded; _Ievent_1 omitted)
i.event|age      _IeveXage_#      (coded as above)
i.event|treat    _IeveXtreat_#    (coded as above)

        failure _d:  censor
  analysis time _t:  time1

Stratified Cox regr. -- Breslow method for ties

No. of subjects =         1296              Number of obs   =       1296
No. of failures =          939
Time at risk    =        39904
                                            Wald chi2(8)    =     103.81
Log likelihood  =   -4152.9183              Prob > chi2     =     0.0000

                        (standard errors adjusted for clustering on id)
------------------------------------------------------------------------
           _t |             Robust
           _d |    Coef.   Std. Err.      z    P>|z|   [95% Conf. Interval]
--------------+---------------------------------------------------------
          age |  .0479377  .0148143    3.24   0.001    .0189022   .0769731
  _IeveXage_2 |  .0051197  .0243052    0.21   0.833   -.0425176   .0527571
  _IeveXage_3 |  .045699   .0279116    1.64   0.102   -.0090068   .1004048
  _IeveXage_4 |  .0049376  .0677345    0.07   0.942   -.1278196   .1376948
        treat |  .664625   .1020419    6.51   0.000    .4646265   .8646234
 _IeveXtrea~2 |  .0146137  .1655469    0.09   0.930   -.3098522   .3390797
 _IeveXtrea~3 | -.7307917  .1799466   -4.06   0.000   -1.08348   -.3781028
 _IeveXtrea~4 | -.9946165  .3253781   -3.06   0.002   -1.632346  -.3568871
------------------------------------------------------------------------
                                                     Stratified by event

. lrtest, saving(0) force
```

```
. xi:stcox age i.event|treat, nohr nolog  noshow strata( event) cluster(id)

i.event          _Ievent_1-4         (naturally coded; _Ievent_1 omitted)
i.event|treat    _IeveXtreat_#       (coded as above)

Stratified Cox regr. -- Breslow method for ties
No. of subjects =         1296                Number of obs   =      1296
No. of failures =          939
Time at risk    =        39904
                                              Wald chi2(5)    =     99.18
Log likelihood  =     -4154.121               Prob > chi2     =    0.0000

                         (standard errors adjusted for clustering on id)
------------------------------------------------------------------------------
          _t |              Robust
          _d |     Coef.   Std. Err.      z    P>|z|    [95% Conf. Interval]
-------------+----------------------------------------------------------------
         age |   .058221   .0102932     5.66   0.000    .0380467   .0783952
       treat |  .6686212   .1023977     6.53   0.000    .4679254   .8693169
_IeveXtrea~2 |  .0131564   .1649829     0.08   0.936   -.3102042    .336517
_IeveXtrea~3 | -.7479445    .180455    -4.14   0.000    -1.10163  -.3942592
_IeveXtrea~4 | -1.000034   .3264192    -3.06   0.002   -1.639804  -.3602641
------------------------------------------------------------------------------
                                                       Stratified by event
. lrtest, force
Cox:  likelihood-ratio test                   chi2(3)       =     2.41
                                              Prob > chi2   =    0.4926
```

We must use the **force** option in the **lrtest** command to perform the test when robust estimation has been used when fitting the models. The likelihood ratio test for the episode by age interaction is not significant indicating that the effect of age does not depend on the number of episode.

(iii) Test if effect of the treatment depend on the episode number

```
. xi:stcox age i.event|treat, nohr strata( event) cluster(id)   issued previously
. lrtest, saving(0) force

. quietly stcox  age treat, nohr strata( event) cluster(id)

. lrtest, force
Cox:  likelihood-ratio test                   chi2(3)       =    25.42
                                              Prob > chi2   =    0.0000
```

The likelihood ratio test for the episode by treatment interaction is highly significant indicating that the effect of treatment is modified by the number of muscle soreness episodes experienced.

Based on the Wald statistics and coefficients estimates, it appears that treatment has a significant effect on the first and second episodes by not so on the third and fourth episodes. This is detailed in the **lincom** output below. The results show that odds of muscle soreness on the standard treatment is about twice that of the new treatment for episodes 1 and 2 and the effect is not significant for episodes 3 and 4.

```
. lincom _b[treat],or

 ( 1)  treat = 0.0

------------------------------------------------------------------------------
          _t |  Odds Ratio   Std. Err.      z    P>|z|     [95% Conf. Interval]
-------------+----------------------------------------------------------------
         (1) |   1.951545    .1998337     6.53   0.000     1.596678    2.385281
------------------------------------------------------------------------------

. lincom _b[treat]+_b[ _IeveXtreat_2],or

 ( 1)  treat + _IeveXtreat_2 = 0.0

------------------------------------------------------------------------------
          _t |  Odds Ratio   Std. Err.      z    P>|z|     [95% Conf. Interval]
-------------+----------------------------------------------------------------
         (1) |   1.977389    .2482199     5.43   0.000     1.546115    2.528965
------------------------------------------------------------------------------

. lincom _b[treat]+_b[ _IeveXtreat_3],or

 ( 1)  treat + _IeveXtreat_3 = 0.0

------------------------------------------------------------------------------
          _t |  Odds Ratio   Std. Err.      z    P>|z|     [95% Conf. Interval]
-------------+----------------------------------------------------------------
         (1) |   .9237412    .1348302    -0.54   0.587     .6939178    1.229681
------------------------------------------------------------------------------

. lincom _b[treat]+_b[ _IeveXtreat_4],or

 ( 1)  treat + _IeveXtreat_4 = 0.0

------------------------------------------------------------------------------
          _t |  Odds Ratio   Std. Err.      z    P>|z|     [95% Conf. Interval]
-------------+----------------------------------------------------------------
         (1) |   .7179087     .223496    -1.06   0.287     .3900113    1.321482
------------------------------------------------------------------------------
```

(c)   Repeat problem 1(a) fitting the conditional B recurrent event model. Does the effect of the covariates depend on the episode number?

The different between the two conditional models is the time scale used. Under Conditional model B we analyze the "gap time" or the length of time from the last to the current episode.

```
. gen gap= time1- time0

. stset gap, failure (censor)

    failure event:  censor ~= 0 & censor ~= .
obs. time interval: (0, gap]
 exit on or before: failure
------------------------------------------------------------------------------
     1296  total obs.
        0  exclusions
------------------------------------------------------------------------------
     1296  obs. remaining, representing
      939  failures in single record/single failure data
    39904  total analysis time at risk, at risk from t =         0
                                earliest observed entry t =         0
                                     last observed exit t =       253
```

(i) Fit of conditional model B

<u>Information matrix based estimates</u>

```
. stcox   age treat, nohr nolog strata(event)  Information matrix-based estimates

         failure _d:  censor
   analysis time _t:  gap

Stratified Cox regr. -- Breslow method for ties

No. of subjects =         1296                Number of obs   =      1296
No. of failures =          939
Time at risk    =        39904
                                              LR chi2(2)      =     69.74
Log likelihood  =    -4583.6185               Prob > chi2     =    0.0000
------------------------------------------------------------------------------
      _t |
      _d |     Coef.   Std. Err.      z    P>|z|    [95% Conf. Interval]
---------+--------------------------------------------------------------------
     age |  .0564612   .0108131     5.22   0.000    .0352678    .0776546
   treat |  .4583486   .0684557     6.70   0.000    .3241779    .5925193
------------------------------------------------------------------------------
                                                       Stratified by event
```

Robust estimates

```
. stcox  age treat, nohr nolog strata(event) cluster(id)   Robust estimates

       failure _d:  censor
   analysis time _t:  gap

Stratified Cox regr. -- Breslow method for ties

No. of subjects   =        1296                Number of obs   =       1296
No. of failures   =         939
Time at risk      =       39904
                                               Wald chi2(2)    =      72.88
Log likelihood    =   -4583.6185               Prob > chi2     =     0.0000

                              (standard errors adjusted for clustering on id)
------------------------------------------------------------------------------
          _t |              Robust
          _d |      Coef.   Std. Err.      z    P>|z|    [95% Conf. Interval]
-------------+----------------------------------------------------------------
         age |   .0564612   .0100427     5.62   0.000    .0367779    .0761445
       treat |   .4583486    .066901     6.85   0.000     .327225    .5894722
------------------------------------------------------------------------------
                                                            Stratified by event
```

Note that the estimates of the standard errors calculated by the two methods are nearly the same.

(ii) Test if effect of treatment and age depend on the episode

```
. xi:stcox i.event|age i.event|treat, nohr nolog strata( event) cluster(id)
i.event            _Ievent_1-4      (naturally coded; _Ievent_1 omitted)
i.event|age        _IeveXage_#      (coded as above)
i.event|treat      _IeveXtreat_#    (coded as above)

         failure _d:  censor
   analysis time _t:  gap

Stratified Cox regr. -- Breslow method for ties
No. of subjects =      1296                  Number of obs   =       1296
No. of failures =       939
Time at risk    =     39904
                                             Wald chi2(8)    =     103.41
Log likelihood  =  -4568.6913                Prob > chi2     =     0.0000

                      (standard errors adjusted for clustering on id)
------------------------------------------------------------------------------
          _t |             Robust
          _d |    Coef.   Std. Err.      z    P>|z|     [95% Conf. Interval]
-------------+----------------------------------------------------------------
         age |  .0479377   .0148143     3.24   0.001     .0189022    .0769731
  _IeveXage_2|  .0042364   .0233278     0.18   0.856    -.0414853    .0499582
  _IeveXage_3|  .0405115   .0296161     1.37   0.171    -.0175349     .098558
  _IeveXage_4|  .0360103   .0678258     0.53   0.595    -.0969258    .1689465
       treat |   .664625   .1020419     6.51   0.000     .4646265    .8646234
 _IeveXtrea~2|  .0018269   .1591581     0.01   0.991    -.3101173     .313771
 _IeveXtrea~3| -.795217    .1801585    -4.41   0.000   -1.148321    -.4421129
 _IeveXtrea~4| -.9028971   .3151975    -2.86   0.004   -1.520673    -.2851214
------------------------------------------------------------------------------
                                                         Stratified by event
```

```
. lrtest, saving(0) force
. xi:stcox age i.event|treat, nohr noshow  nolog strata( event) cluster(id)
i.event          _Ievent_1-4         (naturally coded; _Ievent_1 omitted)
i.event|treat    _IeveXtreat_#       (coded as above)

Stratified Cox regr. -- Breslow method for ties

No. of subjects =        1296                 Number of obs   =      1296
No. of failures =         939
Time at risk    =       39904
                                              Wald chi2(5)    =     99.32
Log likelihood  =   -4569.7758                Prob > chi2     =    0.0000
                       (standard errors adjusted for clustering on id)
------------------------------------------------------------------------------
          _t |              Robust
          _d |    Coef.    Std. Err.       z    P>|z|    [95% Conf. Interval]
-------------+----------------------------------------------------------------
         age |  .0582725    .010209     5.71   0.000    .0382631   .0782818
       treat |  .6686407   .1023719     6.53   0.000    .4679954   .8692859
_IeveXtrea~2 | -.0006665   .1591733    -0.00   0.997   -.3126405   .3113074
_IeveXtrea~3 | -.7970072   .1799742    -4.43   0.000    -1.14975  -.4442642
_IeveXtrea~4 | -.8930505   .3155498    -2.83   0.005   -1.511517  -.2745843
------------------------------------------------------------------------------
                                                         Stratified by event

. lrtest, force
Cox:   likelihood-ratio test                       chi2(3)    =       2.17
                                                   Prob > chi2 =     0.5381
```

(iii) Test if effect of the treatment depend on the episode number

```
. xi:stcox age i.event|treat, nohr strata( event) cluster(id)  issued previously
. lrtest, saving(0) force

. quietly stcox  age treat, nohr strata( event)
    . lrtest, force
     Cox:   likelihood-ratio test                  chi2(3)    =      27.69
                                                   Prob > chi2 =     0.0000
```

The likelihood ratio test for the episode by treatment interaction is highly significant indicating that the effect of treatment is modified by the number of muscle soreness episodes experienced. Based on the Wald statistics and coefficients estimates, it appears that treatment has a significant effect on the first and second episodes by not so on the third and fourth episodes. This is detailed in the **lincom** output below. Treatment is similar for the first and second events and is similar for the third and fourth events. The results show that odds of muscle soreness on the standard treatment are about twice that of the new treatment for episodes 1 and 2 and the effect is not significant for episodes 3 and 4.

```
. lincom _b[treat],or

 ( 1)  treat = 0.0

------------------------------------------------------------------------------
          _t | Odds Ratio   Std. Err.      z    P>|z|     [95% Conf. Interval]
-------------+----------------------------------------------------------------
         (1) |   1.951583    .1997872     6.53  0.000     1.59679    2.385207
------------------------------------------------------------------------------

. lincom _b[treat]+_b[ _IeveXtreat_2],or

 ( 1)  treat + _IeveXtreat_2 = 0.0

------------------------------------------------------------------------------
          _t | Odds Ratio   Std. Err.      z    P>|z|     [95% Conf. Interval]
-------------+----------------------------------------------------------------
         (1) |   1.950282    .2354565     5.53  0.000    1.539332    2.470942
------------------------------------------------------------------------------

. lincom _b[treat]+_b[ _IeveXtreat_3],or

 ( 1)  treat + _IeveXtreat_3 = 0.0

------------------------------------------------------------------------------
          _t | Odds Ratio   Std. Err.      z    P>|z|     [95% Conf. Interval]
-------------+----------------------------------------------------------------
         (1) |    .879531    .1265646    -0.89  0.372    .6633812    1.166109
------------------------------------------------------------------------------

. lincom _b[treat]+_b[ _IeveXtreat_4],or

 ( 1)  treat + _IeveXtreat_4 = 0.0

------------------------------------------------------------------------------
          _t | Odds Ratio   Std. Err.      z    P>|z|     [95% Conf. Interval]
-------------+----------------------------------------------------------------
         (1) |   .7989876    .2378564    -0.75  0.451    .4457996    1.431991
------------------------------------------------------------------------------
```

(d) *Repeat problem 1(a) fitting the marginal recurrent event model. Does the effect of the covariates depend on the episode number?*

Under the marginal recurrent event model each event is considered as a separate process and at risk for all events. As is in the conditional B model, time for each event starts at the beginning of follow-up (0 days) for each subject. The difference is that subjects are considered as censored for the episodes that do not occur.

The most important aspect of analyzing the recurrent event model is the accurate construction of the dataset for analysis (for more detail on data construction, see Table 9.1 Page 309. One must expand the data set to have four observations per-subject. The last record is repeated 4 – EVENT times with CENSOR = 0 if EVENT < 4. In the listing below we see that subject 1 had four events so no additional records are added. However subject 2 had 2 events and was censored at 91 days for the third event. So one more record is added identical to the third and the event is set to 4. One modifies the data for the remainder of the subjects in a similar manner. STATA control language to do this in the current example is listed below.

```
. sort id event
. by id: gen add_rec=cond(_n==_N,4-event+1,.)
. expand add_rec
. sort id event
. by id: replace event = _n
```

```
. stset    time1, failure(censor)

    failure event:  censor ~= 0 & censor ~= .
obs. time interval: (0, time1]
 exit on or before: failure
------------------------------------------------------------------------
     1600  total obs.
        0  exclusions
------------------------------------------------------------------------
     1600  obs. remaining, representing
      939  failures in single record/single failure data
   107044  total analysis time at risk, at risk from t =         0
                                  earliest observed entry t =    0
                                        last observed exit t =  380
```

Note that there are 1600 observations or four per subject.

(i) Fitting the marginal recurrent event model

```
. stcox    age treat, nohr nolog strata(event) cluster(id)    Robust estimates
         failure _d:  censor
    analysis time _t: time1

Stratified Cox regr. -- Breslow method for ties
No. of subjects =             1600                 Number of obs    =     1600
No. of failures =              939
Time at risk    =           107044
                                                   Wald chi2(2)     =    75.21
Log likelihood  =       -4590.9895                 Prob > chi2      =   0.0000

                            (standard errors adjusted for clustering on id)
------------------------------------------------------------------------
             |               Robust
         _t  |
         _d  |   Coef.   Std. Err.      z    P>|z|   [95% Conf. Interval]
-------------+----------------------------------------------------------
         age |  .0728166  .0144297    5.05   0.000    .044535    .1010983
       treat |  .7448138  .0987003    7.55   0.000    .5513648    .9382627
------------------------------------------------------------------------
                                                      Stratified by event
```

(ii) Test if effect of treatment and age depend on the number of episode

```
. xi:stcox i.event|age i.event|treat, nohr nolog strata( event) cluster(id)

i.event          _Ievent_1-4         (naturally coded; _Ievent_1 omitted)
i.event|age      _IeveXage_#         (coded as above)
i.event|treat    _IeveXtreat_#       (coded as above)

         failure _d:  censor
   analysis time _t:  time1

Stratified Cox regr. -- Breslow method for ties

No. of subjects =         1600                  Number of obs   =       1600
No. of failures =          939
Time at risk    =       107044
                                                Wald chi2(8)    =      89.42
Log likelihood  =    -4584.7094                 Prob > chi2     =     0.0000

                            (standard errors adjusted for clustering on id)
------------------------------------------------------------------------------
          _t |                Robust
          _d |      Coef.   Std. Err.      z    P>|z|     [95% Conf. Interval]
-------------+----------------------------------------------------------------
         age |   .0479377   .0148143     3.24   0.001     .0189022    .0769731
  _IeveXage_2|   .0253732    .012819     1.98   0.048     .0002485    .0504979
  _IeveXage_3|   .0723182    .022991     3.15   0.002     .0272566    .1173797
  _IeveXage_4|   .0857806   .0602071     1.42   0.154    -.0322232    .2037844
       treat |    .664625   .1020419     6.51   0.000     .4646265    .8646234
 _IeveXtrea~2|   .2975728   .0975059     3.05   0.002     .1064648    .4886808
 _IeveXtrea~3|  -.0324269   .1570117    -0.21   0.836    -.3401643    .2753104
 _IeveXtrea~4|  -.2052737   .3041472    -0.67   0.500    -.8013914    .3908439
------------------------------------------------------------------------------
                                                          Stratified by event
```

```
. lrtest, saving(0) force

. xi:stcox age i.event|treat, nohr noshow  nolog strata( event) cluster(id)
i.event            _Ievent_1-4         (naturally coded; _Ievent_1 omitted)
i.event|treat      _IeveXtreat_#       (coded as above)

Stratified Cox regr. -- Breslow method for ties

No. of subjects =         1600                Number of obs   =       1600
No. of failures =          939
Time at risk    =       107044
                                              Wald chi2(5)    =      85.06
Log likelihood  =    -4588.292                Prob > chi2     =     0.0000

                              (standard errors adjusted for clustering on id)
------------------------------------------------------------------------------
          _t |               Robust
          _d |     Coef.    Std. Err.      z     P>|z|    [95% Conf. Interval]
-------------+----------------------------------------------------------------
         age |   .0732239   .0145193    5.04   0.000    .0447666    .1016812
       treat |   .6741101   .1034627    6.52   0.000    .4713269    .8768933
_IeveXtrea~2 |   .2880344   .0972154    2.96   0.003    .0974958    .4785731
_IeveXtrea~3 |  -.062402    .1557607   -0.40   0.689   -.3676874    .2428833
_IeveXtrea~4 |  -.2240116   .3055873   -0.73   0.464   -.8229517    .3749284
------------------------------------------------------------------------------
                                                           Stratified by event

. lrtest, force
Cox:  likelihood-ratio test                       chi2(3)      =       7.17
                                                  Prob > chi2  =     0.0668
```

The likelihood ratio test for the episode by age interaction is not significant at the five percent level indicating that the effect of age may not depend on the number of episodes. However it is significant at the 10 percent level. So depending on ones choice of significance we might keep the interaction or not in the model. We proceed deleting it and assess if event modifies the effect of treatment.

(iii) Test if effect of the treatment is modified by episode number.

```
. xi:stcox age i.event|treat, nohr strata( event) cluster(id) issued previously
. lrtest, saving(0) force
. quietly stcox  age treat, nohr nolog strata( event) cluster(id)

. lrtest, force
Cox:  likelihood-ratio test                       chi2(3)      =       5.39
                                                  Prob > chi2  =     0.1451
```

The likelihood ratio test for the episode by treatment interaction is not significant indicating that the effect of treatment does not depend on the number of muscle soreness episodes. Thus the best marginal model is the main effects only model.

```
. stcox age treat, nolog noshow cluster(id) nohr

Cox regression -- Breslow method for ties

No. of subjects =         1600                Number of obs   =      1600
No. of failures =          939
Time at risk    =       107044
                                              Wald chi2(2)    =     71.89
Log likelihood  =   -6143.4628                Prob > chi2     =    0.0000

                       (standard errors adjusted for clustering on id)
------------------------------------------------------------------------------
          _t |               Robust
          _d |     Coef.   Std. Err.      z    P>|z|     [95% Conf. Interval]
-------------+----------------------------------------------------------------
         age |   .055924   .0098311     5.69   0.000     .0366554    .0751926
       treat |  .4336985   .0644314     6.73   0.000     .3074152    .5599817
------------------------------------------------------------------------------

. lincom _b[treat], hr

 ( 1)  treat = 0.0

------------------------------------------------------------------------------
          _t | Haz. Ratio   Std. Err.     z    P>|z|     [95% Conf. Interval]
-------------+----------------------------------------------------------------
         (1) |   1.542954   .0994147    6.73   0.000     1.359906    1.750641
------------------------------------------------------------------------------
```

(e) Prepare a table of estimated hazard ratios, along with 95 percent confidence intervals, comparing the new method to the control method, corresponding to a 10-year change in age, for each of the four models fit. Compare and contrast the point and interval estimates under the four models. Compare and contrast the interpretation of the four sets of point and interval estimates.

Model	Event	HR	Lower B	Upper B
Counting Process	All	1.2723	1.1589	1.3967
Conditional A	1	1.9515	1.5967	2.3853
	2	1.9774	1.5461	2.5289
	3	0.9237	0.6939	1.2297
	4	0.7179	0.3900	1.3215
Conditional B	1	1.9516	1.5968	2.3852
	2	1.9503	1.5393	2.4710
	3	0.8795	0.6634	1.1661
	4	0.7990	0.4458	1.4320
Marginal		1.54295	1.3599	1.75064

The counting process model estimates an overall "average" effect of covariates. The results of conditional models A and B are quite similar and provide event specific estimates of the effect of treatment. As noted earlier the estimate is about 2 for episodes 1 and 2 and is not significant for episodes 3 and 4. The counting process model estimate of 1.15 illustrates the averaging effect of that model. The estimates for the marginal model, 1.5, differ somewhat from the others, but is nearly the average of the 4 estimates from each conditional model.

We did not explore whether or not the episode number modified the estimates of effect for age and treat in the counting process model. A good additional exercise is to perform this analysis. One should find that both age and treat have effects modified by episode number. Using this model provide episode specific estimates of effect for treatment. One should find results similar to models A and B.

2. *One of the classic papers on recurrent event models is Wei, Lin and Weissfeld (1989) (WLW) who propose the marginal method and illustrated its use on recurrence times to bladder cancer. Therneau (1995) presents a detailed discussion of fitting the recurrent events models to the WLW data. These data are available in a library of data sets maintained by the Statistics Department at Carnegie Mellon University. The internet address for the WLW data, Table 2 of Wei, Lin and Weissfeld (1989), is http://lib.stat.cmu.edu/datasets/tumor. Download these data and fit the four recurrent event models using the possibility of four events. Prepare a table of estimated hazard ratios, along with 95 percent confidence intervals. Compare and contrast the interpretation of the four sets of estimates and intervals.*

(i) Codebook

Dataset and its description may be found from http://lib.stat.cmu.edu/datasets/tumor
Data file tomor.dat, obs=86

Tumor Recurrence data for patients with Bladder cancer
Taken from Wei, Lin and Weissfeld, JASA 1989, p 1067.

<u>Variable Name</u>
Group: Treatment group 1 = placebo 2 = thiotepa
FU time: Follow-up times are measured in months.
Number: Initial number of 8 denotes 8 or more initial tumors.
Size: Initial size is measured in centimeters.
rt1:   Time to the 1st recurrence from the beginning of study
rt2:   Time to the 2nd recurrence from the beginning of study
rt3:   Time to the 3rd recurrence from the beginning of study
rt4:   Time to the 4th recurrence from the beginning of study
            Recurrent times are measured in months.
            Up to 4 recurrence times are recorded per patient.

(ii) Data Structure

Group	FUtime	number	size	rt1	rt2	rt3	Rt4
1	1	1	3				
1	4	2	1				
1	7	1	1				
1	10	5	1				
1	10	4	1	6			
1	14	1	1				
1	18	1	1				
1	18	1	3	5			
1	18	1	1	12	16		
1	23	3	3				
1	23	1	3	10	15		
1	23	1	1	3	6	23	
1	24	2	3	7	10	16	24
⋮							
⋮							
2	59	1	3				

These data were collected from a study of 85 subjects randomly assigned to either a treatment group receiving the drug thiotepa or to a group receiving a placebo control. For each patient, time for up to four tumor recurrences was recorded in months.

In the original data, subjects 1 through 4 in the above table, had no tumors recur, thus, each of these 4 patients has only one censored observation spanning from time0 = 0 to end of follow-up (time = ftime) of 1, 4, 7 and 10 months respectively.

Patient 5 in the above table had one tumor recur at 6 months and was followed for 4 more months with follow-up ending at 10 months.

An important aspect of fitting recurrent event models is transforming the raw data into the appropriate structure for the models. For the counting process approach the data file has 178 records of the form shown in Table 9.1. Up to four recurrence times are recorded for each patient. An abbreviated list of some of the key variables in the expanded data set in shown below.

208 CHAPTER 9 SOLUTIONS

```
. list   id group    t0 t1 censor    event

          id       group        t0          t1       censor       event
  1.       1          1          0           1          0            1
  2.       2          1          0           4          0            1
  3.       3          1          0           7          0            1
  4.       4          1          0          10          0            1
  5.       5          1          0           6          1            1
  6.       5          1          6          10          0            2
  7.       6          1          0          14          0            1
  8.       7          1          0          18          0            1
  9.       8          1          0           5          1            1
 10.       8          1          5          18          0            2
  :
  :
171.      82          2          0           4          1            1
172.      82          2          4          24          1            2
173.      82          2         24          47          1            3
174.      82          2         47          50          0            4
175.      83          2          0          54          0            1
176.      84          2          0          38          1            1
177.      84          2         38          54          0            2
178.      85          2          0          59          0            1
```

(i) Counting process model

```
. stset    t1, failure(censor) time0(t0)
       failure event:  censor ~= 0 & censor ~= .
obs. time interval:  (t0, t1]
 exit on or before:  failure
------------------------------------------------------------------------
       178  total obs.
         0  exclusions
------------------------------------------------------------------------
       178  obs. remaining, representing
       112  failures in single record/single failure data
      2480  total analysis time at risk, at risk from t =        0
                                  earliest observed entry t =    0
                                       last observed exit t =   59
```

```
. sum
Variable |      Obs        Mean    Std. Dev.      Min        Max
---------+-----------------------------------------------------------
      id |      178          43     23.47568        1         85
   time1 |      178    7.404494     10.34241        0         47
   time2 |      178    21.33708     14.61479        1         59
  censor |      178    .6292135     .4843779        0          1
   group |      178    1.404494     .4921784        1          2
    size |      178    2.337079     1.825394        1          8
  number |      178    1.960674     1.431535        1          7
   event |      178    1.898876     1.036563        1          4
```

```
. stcox  group size number, nohr nolog cluster(id)

        failure _d:  censor
   analysis time _t:  t1

Cox regression -- Breslow method for ties

No. of subjects =         178                Number of obs   =         178
No. of failures =         112
Time at risk    =        2480
                                             Wald chi2(3)    =       11.62
Log likelihood  =   -453.24263               Prob > chi2     =      0.0088

                        (standard errors adjusted for clustering on id)
------------------------------------------------------------------------------
         _t |                Robust
         _d |     Coef.   Std. Err.      z    P>|z|     [95% Conf. Interval]
------------+-----------------------------------------------------------------
      group | -.4597909   .2595417    -1.77   0.076    -.9684834    .0489015
       size | -.0425622   .0759959    -0.56   0.575    -.1915115    .106387
     number |  .1716441   .061678      2.78   0.005     .0507574    .2925308
------------------------------------------------------------------------------
```

(ii) Conditional Model A

Conditional model assumed a subject is not at risk for a subsequent event until a prior event has occurred and we stratify on the event number.

```
. stcox  group size number, nohr nolog cluster(id) strata( event)

        failure _d:  censor
   analysis time _t:  t1

Stratified Cox regr. -- Breslow method for ties

No. of subjects =         178                Number of obs   =         178
No. of failures =         112
Time at risk    =        2480
                                             Wald chi2(3)    =        7.11
Log likelihood  =   -319.85912               Prob > chi2     =      0.0685

                        (standard errors adjusted for clustering on id)
------------------------------------------------------------------------------
         _t |                Robust
         _d |     Coef.   Std. Err.      z    P>|z|     [95% Conf. Interval]
------------+-----------------------------------------------------------------
      group | -.3342955   .1982339    -1.69   0.092    -.7228268    .0542359
       size | -.0080508   .0604807    -0.13   0.894    -.1265908    .1104892
     number |  .1156526   .0502089     2.30   0.021     .017245     .2140603
------------------------------------------------------------------------------
                                                            Stratified by event
```

(iii) Conditional model B

As noted in the previous problem we analyze the gap time for model B.

```
. gen gap= t1- t0

. stset gap, failure(censor)

     failure event:  censor ~= 0 & censor ~= .
obs. time interval:  (0, gap]
 exit on or before:  failure
--------------------------------------------------------------
      178  total obs.
        0  exclusions
--------------------------------------------------------------
      178  obs. remaining, representing
      112  failures in single record/single failure data
     2480  total analysis time at risk, at risk from t =         0
                             earliest observed entry t =         0
                                  last observed exit t =        59
```

```
. stcox   group size number, nohr nolog cluster(id) strata( event)

         failure _d:  censor
   analysis time _t:  gap


Stratified Cox regr. -- Breslow method for ties
No. of subjects =          178                  Number of obs   =       178
No. of failures =          112
Time at risk    =         2480
                                                Wald chi2(3)    =     11.99
Log likelihood  =   -363.16022                  Prob > chi2     =    0.0074

                               (standard errors adjusted for clustering on id)
------------------------------------------------------------------------------
             |               Robust
       _t   |
       _d   |      Coef.   Std. Err.      z    P>|z|     [95% Conf. Interval]
-------------+----------------------------------------------------------------
      group |  -.2695213   .2093108    -1.29   0.198    -.6797628    .1407203
       size |   .0068402   .0625862     0.11   0.913    -.1158265     .129507
     number |   .1535334   .0491803     3.12   0.002     .0571418    .2499249
------------------------------------------------------------------------------
                                                         Stratified by event
```

(iv) Marginal model

As noted in the previous problem, to fit the marginal model we need to expand the data set so that each subject has four records. A brief list of the expanded data set is shown below. Stata code to do this is listed in problem 1.

```
. list  id group   t0 t1 censor  event
              id      group        t0            t1         censor       event
   1.        1         1           0             1            0            1
   2.        1         1           0             1            0            2
   3.        1         1           0             1            0            3
   4.        1         1           0             1            0            4
   5.        2         1           0             4            0            1
   6.        2         1           0             4            0            2
   7.        2         1           0             4            0            3
   8.        2         1           0             4            0            4
   :
   :
 333.       84         2           0            38            1            1
 334.       84         2          38            54            0            2
 335.       84         2          38            54            0            3
 336.       84         2          38            54            0            4
 337.       85         2           0            59            0            1
 338.       85         2           0            59            0            2
 339.       85         2           0            59            0            3
 340.       85         2           0            59            0            4
```

```
. stset  t1, failure( censor)

     failure event:  censor ~= 0 & censor ~= .
obs. time interval:  (0, t1]
 exit on or before:  failure

------------------------------------------------------------------------
       340  total obs.
         0  exclusions
------------------------------------------------------------------------
       340  obs. remaining, representing
       112  failures in single record/single failure data
      8522  total analysis time at risk, at risk from t =         0
                                   earliest observed entry t =    0
                                        last observed exit t =   59
```

```
. stcox   group size number, nohr nolog cluster(id) strata( event)

        failure _d:  censor
   analysis time _t: t1

Stratified Cox regr. -- Breslow method for ties

No. of subjects =         340                   Number of obs   =        340
No. of failures =         112
Time at risk    =        8522
                                                Wald chi2(3)    =      15.38
Log likelihood  =   -428.05791                  Prob > chi2     =     0.0015

                           (standard errors adjusted for clustering on id)
------------------------------------------------------------------------------
          _t |                 Robust
          _d |      Coef.   Std. Err.      z    P>|z|     [95% Conf. Interval]
-------------+----------------------------------------------------------------
       group |  -.5798608   .3052361    -1.90   0.057    -1.178113    .018391
        size |  -.0509387   .0935879    -0.54   0.586    -.2343675   .1324902
      number |   .2084914   .0660645     3.16   0.002     .0790073   .3379755
------------------------------------------------------------------------------
                                                           Stratified by event
```

Four sets of point and interval estimates.

Model	Variable	HR	LB	UB
Counting Process	RX	0.63	0.38	1.05
	Number	1.19	1.05	1.34
	Size	0.99	0.83	1.11
Conditional A	RX	0.72	0.49	1.06
	Number	1.12	1.02	1.24
	Size	0.99	0.88	1.12
Conditional A	RX	0.76	0.51	1.15
	Number	1.17	1.06	1.28
	Size	1.01	0.89	1.14
Marginal	RX	0.56	0.31	1.02
	Number	1.23	1.08	1.4
	Size	0.95	0.79	1.14

The estimates of effect under the four different models are nearly identical. Treatment and size are not significant and the HR for number is approximately 1.2 in all four models.

3. *In the WHAS a reasonable proportional hazards model, when the analysis is restricted to grouped cohort one (YRGRP =1), contains age and left heart failure complications (CHF). Use the methods for including a subject-specific unobserved frailty to explore possible unaccounted for heterogeneity among the subjects in grouped cohort one.*

A frailty model includes the value of an additional unmeasured covariate (frailty) in the hazard function. Since hazard cannot be negative, distribution of frailty must have positive values. The most frequently used distribution of frailty is gamma with mean equal to one and variance parameter θ. If the value of frailty is greater than one, the subject has a larger than average hazard and is said to be more frail.

The major problem faced by the practitioner wishing to use any frailty model is the lack of readily available software. The **streg** procedure in Stata 7 can now estimate parametric survival models with frailty, but not the proportional hazards model. Two forms of the frailty distribution allowed are gamma and inverse Gaussian. A frailty is allowed with all the parametric distributions currently available in **streg**.

The proportional hazards frailty models fit in the text use complicated STATA programs we wrote. These programs are tricky to use and are quite time consuming. Hence we approach the fitting problem by using the frailty option available in **streg.**

Some preliminary analyses show that the Weibull model fits much better than the exponential model. The fit of the basic Weibull in hazard form is shown below. Note that the test for shape parameter in the form $1/\ln(p)$ is significant.

```
. streg   age chf, nohr nolog noshow dist(weib) nohr

Weibull regression -- log relative-hazard form

No. of subjects =            160                Number of obs   =        160
No. of failures =            102
Time at risk    =         415690
                                                LR chi2(2)      =      23.69
Log likelihood  =     -317.94324                Prob > chi2     =     0.0000

------------------------------------------------------------------------------
          _t |      Coef.   Std. Err.       z    P>|z|     [95% Conf. Interval]
-------------+----------------------------------------------------------------
         age |   .0343165   .0090967     3.77    0.000     .0164873    .0521456
         chf |   .3426082   .2098048     1.63    0.102    -.0686016    .7538181
       _cons |  -5.714671   .6881453    -8.30    0.000    -7.063411   -4.365931
-------------+----------------------------------------------------------------
       /ln_p |  -.948282    .0870254   -10.90    0.000    -1.118849   -.7777153
-------------+----------------------------------------------------------------
           p |   .387406    .0337142                       .3266557    .4594545
         1/p |  2.581271    .2246362                      2.176494    3.061328
------------------------------------------------------------------------------
```

The results of fitting the same model with a gamma distribution frailty are shown below

```
. streg  age chf, nohr nlog noshow dist(weib) nohr frailty(gamma)

Weibull regression -- log relative-hazard form
                    Gamma frailty

No. of subjects =          160                  Number of obs    =       160
No. of failures =          102
Time at risk    =       415690
                                                LR chi2(3)       =     24.67
Log likelihood  =   -317.45414                  Prob > chi2      =    0.0000

------------------------------------------------------------------------------
        _t |      Coef.   Std. Err.      z    P>|z|     [95% Conf. Interval]
-----------+------------------------------------------------------------------
       age |   .044426    .0157706     2.82   0.005     .0135163    .0753358
       chf |   .6079699   .4410166     1.38   0.168    -.2564067    1.472347
     _cons |  -6.696069   1.387204    -4.83   0.000    -9.414939   -3.977199
-----------+------------------------------------------------------------------
     /ln_p |  -.7891515   .1996438    -3.95   0.000    -1.180446   -.3978568
    /ln_the|  -.3756754   1.228541    -0.31   0.760    -2.783571    2.032221
-----------+------------------------------------------------------------------
         p |   .4542301   .0906842                      .3071417    .6717582
       1/p |   2.201528   .4395214                      1.488631    3.255827
     theta |   .6868253   .843793                       .0618173    7.631014
------------------------------------------------------------------------------
Likelihood ratio test of theta=0: chibar2(01) =    0.98 Prob>=chibar2 = 0.161
```

The gamma frailty parameter, theta, is not significant, which indicates that we cannot say that there is unaccounted for heterogeneity in the data. The estimates of the age coefficient are similar but the estimate of the chf coefficient and its standard error are larger under the frailty model.

4. *In order to provide a data set for applying methods for the analysis of a nested case-control study, we created a small data set from the main WHAS data (<u>this is not a subset of the WHAS data described in Section 1.3</u>). Before performing the case-control sampling we broke ties in survival times by subtracting a uniform (0,1) random variable from each value of LENFOL corresponding to a death, and censored values were not changed. The modified version of LENFOL is denoted as T. The sampling procedure selected five controls for each case. These data are in the file WHASNCC.DAT that may be found on the web sites containing the data sets from this text. The variables are (see Table 1.4 for a description) SET, CASE, T, LENFOL, FSTAT, AGE, SEX, CHF, MIORD, YRGRP, LENSTAY and NR. The variable describing the set of one case and five controls is SET where CASE = 1 for the indexed death time and NR is the number of subjects in the risk set in the original cohort.*

The data set on the web was created from the original Worcester Heart Attack Study data and should be used for the analyses in this problem.

In order to illustrate the method of creating a nested case-control study we apply the **sttooc** command to the WHAS data set used in this next.

Method used to create a nested case-control study.

In some follow-up studies the number of subjects too large. The nested case-control study offers the possibility of fitting data from a follow-up study using fewer subjects than the entire cohort. If an adequate number of controls are used, the results form the nested study should agree with those that would have been obtained from entire cohort.

The **sttocc** procedure in Stata can sample and create an output data set containing all the information needed to fit the model using the methods described in the text based on conditional logistic regression program. We used the **sttocc** procedure to create WHASNCC.DAT file.

```
. gen t= lenfol
. set seed 254587
. replace t=t-0.01*uniform() if  fstat==1
(249 real changes made)

. stset t fstat

     failure event:  fstat ~= 0 & fstat ~= .
obs. time interval:  (0, t]
 exit on or before:  failure
------------------------------------------------------------------------
       481  total obs.
         0  exclusions
------------------------------------------------------------------------
       481  obs. remaining, representing
       249  failures in single record/single failure data
    834553.7  total analysis time at risk, at risk from t =         0
                            earliest observed entry t =         0
                             last observed exit t =          5843
. sts gen nr=n
. sttocc,n(5) nodots

        failure _d:  fstat
   analysis time _t:  t
There are 249 cases
Sampling 5 controls for each case

note
The sttocc procedure automatically generates 3 variables
_case : a case-control indicator coded 0 for controls and 1 for cases
 set  : a case-control set identifier (_set)
 time : the time, on the analysis scale at which the set was constructed, that
        is the failure time of the case
```

As noted this data set has been created to illustrate the method.

216 *CHAPTER 9 SOLUTIONS*

(a) *Use the methods for the analysis of nested case-control studies to fit the proportional hazards model to these data.*

The model contains only one continuous covariate. Rather than begin by fitting the base model and then checking for the scale of age we move directly to using **fracpoly**.

```
. fracpoly clogit case age sex chf miord, group( set) comp
........
-> gen double Iage__1 = X^-1-.1519 if e(sample)
-> gen double Iage__2 = X^-1*ln(X)-.2863 if e(sample)
   (where: X = age/10)

Iteration 0:   log likelihood = -434.09113
Iteration 1:   log likelihood =  -399.3013
Iteration 2:   log likelihood = -399.14753
Iteration 3:   log likelihood = -399.14743

Conditional (fixed-effects) logistic regression   Number of obs   =       1494
                                                  LR chi2(5)      =      94.00
                                                  Prob > chi2     =     0.0000
Log likelihood = -399.14743                       Pseudo R2       =     0.1053

------------------------------------------------------------------------------
        case |      Coef.   Std. Err.      z    P>|z|     [95% Conf. Interval]
-------------+----------------------------------------------------------------
     Iage__1 |    14.8491   5.481206     2.71   0.007     4.106129    25.59206
     Iage__2 |  -37.55501   8.235452    -4.56   0.000     -53.6962   -21.41382
         sex |   .0443711   .1510991     0.29   0.769    -.2517776    .3405199
         chf |   .6428418   .1514358     4.24   0.000     .3460331    .9396505
       miord |    .429864   .1495133     2.88   0.004     .1368234    .7229047
------------------------------------------------------------------------------
Deviance: 798.295. Best powers of age among 44 models fit: -1 -1.

Fractional polynomial model comparisons:
------------------------------------------------------------
age              df      Deviance   Gain    P(term)  Powers
------------------------------------------------------------
Not in model      0       835.239    --       --
Linear            1       803.621   0.000    0.000    1
m = 1             2       802.895   0.727    0.394    2
m = 2             4       798.295   5.327    0.100   -1 -1
------------------------------------------------------------
```

The p-value for the best $m = 2$ vs. $m = 1$ is 0.100 (NS) so we proceed treating age as linear.

```
. clogit _case age sex chf miord, group( _set) nolog

Conditional (fixed-effects) logistic regression   Number of obs   =     1494
                                                  LR chi2(4)      =    88.67
                                                  Prob > chi2     =   0.0000
Log likelihood = -401.8107                        Pseudo R2       =   0.0994

------------------------------------------------------------------------------
        case |      Coef.   Std. Err.      z    P>|z|     [95% Conf. Interval]
-------------+----------------------------------------------------------------
         age |   .0360374   .0066103     5.45   0.000     .0230814    .0489934
         sex |   .0538552   .1506091     0.36   0.721    -.2413333    .3490437
         chf |   .6575245    .150909     4.36   0.000     .3617484    .9533006
       miord |   .4092602   .1482066     2.76   0.006     .1187805    .6997398
------------------------------------------------------------------------------
```

(b) *Following the fit of the model in problem 4(a) prepare a table of estimated hazard ratios with corresponding 95 percent confidence intervals.*

The results from **lincom** can be used for the table.

```
. lincom 10*_b[age],or

 ( 1)  10.0 age = 0.0

------------------------------------------------------------------------------
        case | Odds Ratio   Std. Err.      z    P>|z|     [95% Conf. Interval]
-------------+----------------------------------------------------------------
         (1) |   1.433865   .0947833     5.45   0.000     1.259624    1.632208
------------------------------------------------------------------------------

. lincom _b[sex],or

 ( 1)  sex = 0.0

------------------------------------------------------------------------------
        case | Odds Ratio   Std. Err.      z    P>|z|     [95% Conf. Interval]
-------------+----------------------------------------------------------------
         (1) |   1.055332   .1589426     0.36   0.721     .7855797    1.417711
------------------------------------------------------------------------------

. lincom _b[chf],or

 ( 1)  chf = 0.0

------------------------------------------------------------------------------
        case | Odds Ratio   Std. Err.      z    P>|z|     [95% Conf. Interval]
-------------+----------------------------------------------------------------
         (1) |   1.930009   .2912556     4.36   0.000     1.435838    2.594258
------------------------------------------------------------------------------
```

```
. lincom _b[miord],or

 ( 1)  miord = 0.0

------------------------------------------------------------------------------
        case | Odds Ratio   Std. Err.      z    P>|z|     [95% Conf. Interval]
-------------+----------------------------------------------------------------
         (1) |   1.505703    .2231552     2.76   0.006     1.126123    2.013229
------------------------------------------------------------------------------
```

(c) *Graph the estimated covariate-adjusted survivorship functions comparing the survival experience of those with and without left heart complications, that is, CHF = 0 vs. CHF = 1.*

```
. predict xb,xb
. gen theta=exp(xb)
. sort _set _case
. quietly by _set:generate sumth=cond(_n==_N,sum(theta),.)
. gen w= nr/6
. gen h0_t=1/(w*sumth)
. sort t
. gen H0_t=cond(case==1,sum(h0_t),.)
. gen rm= xb-_b[chf]
. sum rm, det
. gen S0_t=exp(- H0_t)
. gen S_0= S0_t^exp(2.033484)
. gen S_1= S0_t^exp(2.033484+_b[chf])
```

```
. graph S_0 S_1 t, s(ii) c(JJ) yscale(0,1) ylab(0, .25, .5, .75, 1) xscale(0,4600)
```

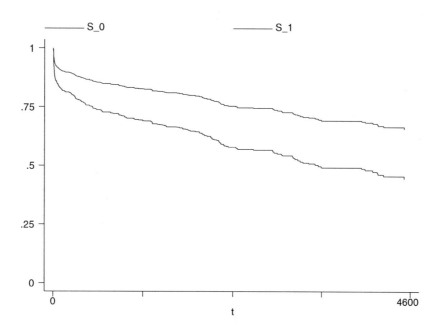

5. *Review the discussion in Section 9.5 of the plots of the estimated cumulative regression coefficients obtained from fitting Aalen's additive model to the UIS data.*

Based on the plots and the discussion in Chapter 9 several time varying effects were postulated. It appeared that Beck score might have an early effect that disappeared after 200 days. Thus as a function of time we wish to test the following hazard function in Beck score (x):

$$h(t,\beta,x) = \beta_1 x \times (1-d_1(t)) + \beta_2 x \times d_1(t) = \begin{cases} \beta_1 x \text{ if } t < 200 \\ \beta_2 x \text{ if } t \geq 200 \end{cases}$$

where

$$d_1(t) = \begin{cases} 0 \text{ if } t < 200 \\ 1 \text{ if } t \geq 200 \end{cases}$$

The test for no late effect is $H_o: \beta_2 = 0$.

The covariate ivhx_3 measures recent history of IV drug use and it appears from figure 9.2(e) that it may have no effect up to 350 days. The hazard function for this effect is

$$h(t,\beta,x) = \beta_1 x \times (1-d_2(t)) + \beta_2 x \times d_2(t) = \begin{cases} \beta_1 x \text{ if } t < 350 \\ \beta_2 x \text{ if } t \geq 350 \end{cases}$$

where

$$d_2(t) = \begin{cases} 0 & \text{if } t < 350 \\ 1 & \text{if } t \geq 350 \end{cases}.$$

The test for no early effect is $H_0: \beta_1 = 0$.

The covariate race appears from figure 9.2(f) to have only an effect up to about 150 days. The hazard function for this model is

$$h(t, \beta, x) = \beta_1 x \times (1 - d_3(t)) + \beta_2 x \times d_3(t) = \begin{cases} \beta_1 x & \text{if } t < 150 \\ \beta_2 x & \text{if } t \geq 150 \end{cases}$$

where

$$d_3(t) = \begin{cases} 0 & \text{if } t < 150 \\ 1 & \text{if } t \geq 150 \end{cases}.$$

The test for no late effect is $H_0: \beta_2 = 0$.

The model appears in figure 9.2(g) to be a little more complicated in treatment with different effects in three intervals of time. In this case we need an expanded model as follows

$$h(t, \beta, x) = \beta_1 x \times d_4\_1(t) + \beta_2 x \times d_4\_2(t) + \beta_3 d_4\_3(t) = \begin{cases} \beta_1 x & \text{if } t < 90 \\ \beta_2 x & \text{if } 90 \leq t \leq 180 \\ \beta_3 x & \text{if } t > 180 \end{cases}$$

where

$$d_4\_k(t) = \begin{cases} 1 & \text{if } d_4(t) = k \\ 0 & \text{otherwise} \end{cases} \quad \text{and} \quad d_4(t) = \begin{cases} 1 & \text{if } t < 90 \\ 2 & \text{if } 90 \leq t \leq 180 \\ 3 & \text{if } t \geq 180 \end{cases}.$$

The test for no early effect in interval j is $H_0: \beta_j = 0, j = 1, 2, 3$.

As a result we want to fit the following proportional hazards model in the main effects from the UIS Study:

$$h(t, \mathbf{x}, \boldsymbol{\beta}) = h_0(t) \times \begin{bmatrix} \beta_1 age + \beta_2 ndrugtx + \beta_3 site \\ + \beta_4 becktota \times (1 - d_1(t)) + \beta_5 becktota \times d_1(t) \\ + \beta_6 ivhx\_3 \times (1 - d_2(t)) + \beta_7 ivhx\_3 \times d_2(t) \\ + \beta_8 race \times (1 - d_3(t)) + \beta_9 race \times d_3(t) \\ \beta_{10} x \times d_4\_1(t) + \beta_{11} x \times d_4\_2(t) + \beta_{12} d_4\_3(t) \end{bmatrix}.$$

We fit this model in problem 5(a).

(a) *Fit a proportional hazards model using the time-varying covariates suggested in the plots. Assess the need to include additional interactions. Compare this new "final" model to the previously obtained final model shown in Tables 5.11 and 6.6. Which model provides the better description of the effect of the covariates? Consider this question from both statistical and clinical perspectives.*

The first step in fitting the above model is to create the three time varying dichotomous covariates. Before doing this we fit the base model with no time varying covariates.

```
. stcox age ndrugtx treat site becktota  ivhx_3 race, nohr nolog noshow

Cox regression -- Breslow method for ties

No. of subjects =          575                Number of obs   =      575
No. of failures =          464
Time at risk    =       138900
                                              LR chi2(7)      =    44.51
Log likelihood  =   -2641.7294                Prob > chi2     =   0.0000

------------------------------------------------------------------------------
          _t |
          _d |      Coef.   Std. Err.      z    P>|z|     [95% Conf. Interval]
-------------+----------------------------------------------------------------
         age |  -.0261517   .0080491    -3.25   0.001    -.0419276   -.0103758
     ndrugtx |   .0290737   .0082126     3.54   0.000     .0129773     .04517
       treat |  -.2324266   .0937326    -2.48   0.013    -.4161392   -.048714
        site |  -.0866855   .1078637    -0.80   0.422    -.2980944   .1247233
    becktota |   .0083976   .0049516     1.70   0.090    -.0013074   .0181025
      ivhx_3 |   .2561209   .1062996     2.41   0.016     .0477776   .4644642
        race |   -.224462   .1152656    -1.95   0.051    -.4503784   .0014544
------------------------------------------------------------------------------
```

The next step is to use the **stsplit** command to create the appropriate time intervals where the dichotomous covariates are constant.

```
. stsplit d4,at(90 180)
(691 observations (episodes) created)

. tab d4, gen(d4_) * This generates the 3  0-1 design variables for each
interval.

         d4 |      Freq.     Percent        Cum.
------------+-----------------------------------
          0 |        575       45.42       45.42
         90 |        414       32.70       78.12
        180 |        277       21.88      100.00
------------+-----------------------------------
      Total |       1266      100.00

. stsplit   d1,at(200)
(253 observations (episodes) created)
```

```
. tab d1

          d1 |      Freq.     Percent        Cum.
-------------+-----------------------------------
           0 |       1266       83.34       83.34
         200 |        253       16.66      100.00
-------------+-----------------------------------
       Total |       1519      100.00

. replace d1=1 if d1==200
(253 real changes made)

. stsplit  d2,at(350)
(156 observations (episodes) created)

. tab d2

          d2 |      Freq.     Percent        Cum.
-------------+-----------------------------------
           0 |       1519       90.69       90.69
         350 |        156        9.31      100.00
-------------+-----------------------------------
       Total |       1675      100.00

. replace d2=1 if d2==350
(156 real changes made)

. stsplit d3,at(150)
(318 observations (episodes) created)

. tab d3

          d3 |      Freq.     Percent        Cum.
-------------+-----------------------------------
           0 |        989       49.62       49.62
         150 |       1004       50.38      100.00
-------------+-----------------------------------
       Total |       1993      100.00

. replace d3=1 if d3==150
(1004 real changes made)
```

The first thing we do is fit the base model to be sure we get the same estimates from the record split data and as shown we do.

```
. stcox age ndrugtx treat site becktota  ivhx_3 race, nohr nolog noshow

Cox regression -- Breslow method for ties

No. of subjects  =            575              Number of obs   =        1993
No. of failures  =            464
Time at risk     =         138900
                                                LR chi2(7)      =       44.51
Log likelihood   =     -2641.7294               Prob > chi2     =      0.0000

------------------------------------------------------------------------------
          _t |
          _d |      Coef.   Std. Err.       z     P>|z|    [95% Conf. Interval]
-------------+----------------------------------------------------------------
         age |  -.0261517   .0080491    -3.25    0.001    -.0419276   -.0103758
      ndrugtx|   .0290737   .0082126     3.54    0.000     .0129773     .04517
       treat |  -.2324266   .0937326    -2.48    0.013    -.4161392   -.048714
        site |  -.0866855   .1078637    -0.80    0.422    -.2980944    .1247233
     becktota|   .0083976   .0049516     1.70    0.090    -.0013074    .0181025
      ivhx_3 |   .2561209   .1062996     2.41    0.016     .0477776    .4644642
        race |   -.224462   .1152656    -1.95    0.051    -.4503784    .0014544
------------------------------------------------------------------------------
```

Next we add to the model the needed interaction terms in coefficients. To do this we first need to generate the interactions.

```
. gen beckearly=becktota*(1-d1)

. gen becklate=becktota*(d1)

. gen iv3early=ivhx_3*(1-d2)

. gen iv3late=ivhx_3*d2

. gen raceearly=race*(1-d3)

. gen racelate=race*d3

. gen treat_int1=treat*d4_1

. gen treat_int2=treat*d4_2

. gen treat_int3=treat*d4_3
```

```
. stcox age ndrugtx  site beckearly becklate   iv3early  iv3late  raceearly
racelate treat_int1 trea t_int2 treat_int3, nohr nolog noshow

Cox regression -- Breslow method for ties

No. of subjects =         575                Number of obs    =      1993
No. of failures =         464
Time at risk    =      138900
                                             LR chi2(12)      =     57.80
Log likelihood  =   -2635.0838               Prob > chi2      =    0.0000

------------------------------------------------------------------------------
         _t |
         _d |      Coef.   Std. Err.      z    P>|z|     [95% Conf. Interval]
------------+-----------------------------------------------------------------
        age |  -.0265378   .0080507    -3.30   0.001    -.0423168   -.0107587
     ndrugtx|   .0292854   .0082007     3.57   0.000     .0132124    .0453584
       site |  -.0860812   .1081297    -0.80   0.426    -.2980115    .1258492
   beckearly|   .0157173   .0058924     2.67   0.008     .0041685    .0272661
    becklate|  -.0071591    .009178    -0.78   0.435    -.0251477    .0108295
    iv3early|   .2130233   .1105908     1.93   0.054    -.0037307    .4297772
     iv3late|   .6371051   .3055048     2.09   0.037     .0383268    1.235883
   raceearly|  -.3344326   .1588109    -2.11   0.035    -.6456962   -.0231691
    racelate|   -.119297   .1616299    -0.74   0.460    -.4360858    .1974918
  treat_int1|  -.2612087   .1592299    -1.64   0.101    -.5732934    .0508761
  treat_int2|  -.5416687   .1736115    -3.12   0.002     -.881941   -.2013964
  treat_int3|   .0331717   .1580339     0.21   0.834    -.2765691    .3429125
------------------------------------------------------------------------------
```

Looking at the *p* - values of the Wald tests we see that:
1. For Beck score we reject the hypothesis of no early effect and fail to reject the hypothesis of no late effect. This supports what we saw in the Figure 9.2(c).
2. For ivhx_3 we fail to reject (barely as $p = 0.054$) the hypothesis of no early effect and reject the hypothesis of no late effect. This tends to support a model with modest early effect and a significant late effect in ivhx_3.
3. For Beck score we reject the hypothesis of no early effect and fail to reject the hypothesis of no late effect. This supports what we saw in the Figure 9.2(c).
4. For race score we reject the hypothesis of no early effect and fail to reject the hypothesis of no late effect. This supports what we saw in the Figure 9.2(f).
5. For treatment we fail to reject the hypothesis of no effect in the first 90 days, reject the hypothesis of no effect between 90 and 180 days and fail to reject the hypothesis of no effect after 180 days. This supports what we saw in the Figure 9.2(g).

Thus the use of the Aalen plots and corresponding time varying covariates has identified important time varying effects in the covariates. The next step is to replace ndrugtx with its fractional polynomials and add the interactions of age and site and race and site.

```
. xi:stcox age  ndrugfp1 ndrugfp2   site  beckearly  becklate   iv3early  iv3late
treat_int1 treat_int2 treat_int3 raceearly racelate   i.raceearly|site
i.racelate|site agexsite , nohr nolog noshow

i.raceearly         _Iraceearly_0-1      (naturally coded; _Iraceearly_0 omitted)
i.raceea~y|site     _IracXsite_#         (coded as above)
i.racelate          _Iracelate_0-1       (naturally coded; _Iracelate_0 omitted)
i.racelate|site     _IracXsitea#         (coded as above)
note: site dropped due to collinearity

Cox regression -- Breslow method for ties
No. of subjects =         575                  Number of obs    =        1993
No. of failures =         464
Time at risk    =      138900
                                                LR chi2(16)     =       81.51
Log likelihood  =   -2623.2288                  Prob > chi2     =      0.0000
```

_t _d	Coef.	Std. Err.	z	P>|z|	[95% Conf. Interval]	
age	-.0418229	.0098837	-4.23	0.000	-.0611947	-.0224512
ndrugfp1	-.5825523	.1254805	-4.64	0.000	-.8284895	-.3366151
ndrugfp2	-.2176667	.0486919	-4.47	0.000	-.313101	-.1222324
beckearly	.0164704	.005885	2.80	0.005	.0049361	.0280048
becklate	-.0078665	.0092327	-0.85	0.394	-.0259622	.0102292
iv3early	.1853061	.1127123	1.64	0.100	-.0356059	.4062181
iv3late	.6192882	.3072786	2.02	0.044	.0170331	1.221543
treat_int1	-.2766709	.1596091	-1.73	0.083	-.5894991	.0361573
treat_int2	-.5670863	.1740344	-3.26	0.001	-.9081874	-.2259852
treat_int3	.0269035	.1585277	0.17	0.865	-.2838051	.3376121
raceearly	-.6086433	.1885812	-3.23	0.001	-.9782556	-.239031
racelate	-.3414249	.1820557	-1.88	0.061	-.6982475	.0153976
site	-1.327751	.5314015	-2.50	0.012	-2.369279	-.2862231
_IracXsite_1	.9516651	.3198475	2.98	0.003	.3247755	1.578555
_IracXsitea1	.8024474	.3408708	2.35	0.019	.1343528	1.470542
agexsite	.0326358	.0160741	2.03	0.042	.0011311	.0641406

The above results show that the p-values for the Wald tests for the three interactions and the fractional polynomials are significant. The remaining variables retain their respective significance or lack there of.

Thus we make two points here:
1. The Aalen additive model has provided a graphical basis for identifying time varying effect of covariates in the PH model.
2. The addition of time varying covariate to the main UIS model from Chapter 6 has improved/sharpened the estimates of effect and how they depend on time.

The procedure for estimating hazard ratios is the same as that illustrated in detailed in Chapter 6. The difference here is that for the "xearly" "xlate" and "treat_intx" variables the hazard ratios apply only to the relevant interval of time

226 *CHAPTER 9 SOLUTIONS*

(b) *Use plots from the estimated cumulative regression coefficients from a fit of the Aalen additive model to explore the possibility of a time-varying effect in NDRUGFP1 and NDRUGFP2.*

Note: We use a STATA program we wrote to fit the Aalen model. The program is rather lengthy so we do not list it here. Any reader who would like a copy should contact us.

We begin by centering the data and setting up the data.

```
. sum age

Variable |     Obs        Mean    Std. Dev.       Min         Max
---------+-----------------------------------------------------
     age |     575    32.38261    6.193149         20          56

. gen age_c=age-32.38261

. sum becktota

Variable |     Obs        Mean    Std. Dev.       Min         Max
---------+-----------------------------------------------------
becktota |     575    17.36743    9.332962          0          54

. gen beck_c=becktota-17.36743

. sum  ndrugfp1

Variable |     Obs        Mean    Std. Dev.       Min         Max
---------+-----------------------------------------------------
ndrugfp1 |     575    3.550733    2.913425    .2439024          10

. gen ndgfp1_c= ndrugfp1-3.550753

. sum ndrugfp2

Variable |     Obs        Mean    Std. Dev.       Min         Max
---------+-----------------------------------------------------
ndrugfp2 |     575   -5.548602    7.487675   -23.02585     .367871

. gen ndgfp2_c= ndrugfp2+5.548602

. sum   age_c beck_c ndgfp1_c ndgfp2_c

Variable |     Obs        Mean    Std. Dev.       Min         Max
---------+-----------------------------------------------------
   age_c |     575   -1.33e-06    6.193149   -12.38261    23.61739
  beck_c |     575   -2.12e-06    9.332962   -17.36743    36.63257
ndgfp1_c |     575    -.0000203    2.913425   -3.306851    6.449247
ndgfp2_c |     575   -5.08e-08    7.487676   -17.47725    5.916473
```

```
. gen t=time

. set seed 12345

. replace t=t-0.01*uniform() if censor==1 * we have to break the ties to run the Aalen
model*
(464 real changes made)

. aalen t censor  age_c beck_c ndgfp1_c ndgfp2_c ivhx_3 race treat site

 Fit of Aalen's Additive Model
 Using  t  as the time variable and censor as  the censoring variable

 Model variables are: age_c beck_c ndgfp1_c ndgfp2_c ivhx_3 race treat site

Output omitted.
```

Since we are only interested in ndrugfp1 and ndrugfp2, we look at there graphs.

The figure below is for ndrugfp1

```
. graph using aA4
```

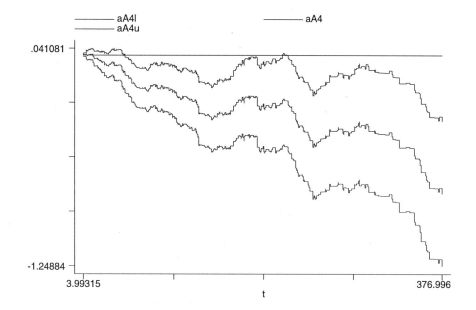

The figure below is for ndrugfp2

. **graph using aA5**

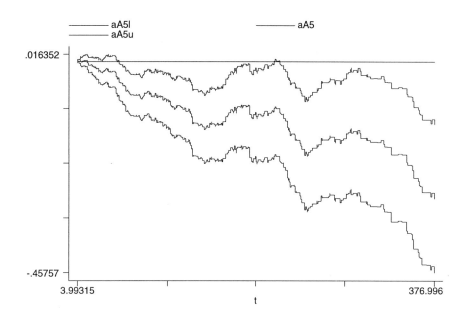

The two graphs shown above generally decrease in a linear fashion over the entire range of time, supporting the conclusion that the model is linear in the transformed variables for number of drug treatments. The fact that the two graphs look quite similar is due to the high correlation between the two transformed variables.

6. *Fit the Aalen model to the WHAS data using a model containing AGE, SEX, CHF and MIORD. Use plots from the estimated cumulative regression coefficients from a fit of the Aalen additive model to explore the possibility of a time-varying effect in AGE, SEX, CHF and MIORD. Fit the proportional hazards model containing AGE, SEX, CHF and MIORD. Fit the proportional hazards model including any time-varying covariate effects suggested by the plots of the estimated cumulative regression coefficients from the fit of the Aalen model. Are the time-varying effects significant and does their inclusion improve the model from a statistical and clinical perspective?*

```
. gen t= lenfol
. replace t= t-0.01*uniform() if fstat==1
(249 real changes made)
. sum age
. gen age_c=age - r(mean)
. aalen  t fstat   age_c sex chf miord01
   *output omitted
```

The figure below shows the Aalen plots the cumulative regression coefficients for the four covariates age, sex, chf and miord (coded 0,1).

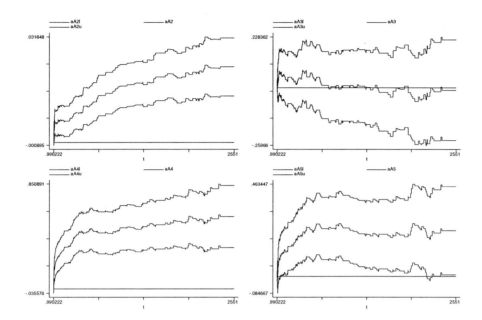

In the above plot the graph for age (top left) increases linearly over the entire range and thus shows no time varying effect. The graph for sex (top right) has constant nearly zero slopes and the confidence bands enclose the zero line and we conclude sex has no effect. The graph for chf (bottom left) shows a sharp significant linear increase in the first about 600 days and then still increases linearly but with much less slope. This graph suggests that chf has a time varying effect that changes at about 600 days. We will try modeling it in the PH setting with

$$h(t,\beta,x) = \beta_1 x \times (1 - d_1(t)) + \beta_2 x \times d_1(t) = \begin{cases} \beta_1 x \text{ if } t < 600 \\ \beta_2 x \text{ if } t \geq 600 \end{cases}$$

where

$$d_1(t) = \begin{cases} 0 \text{ if } t < 600 \\ 1 \text{ if } t \geq 600 \end{cases}.$$

The two slope coefficients correspond to the time varying effects (log-hazard scale). The same general pattern is present in the graph for miord (bottom right). The difference is that after 600 days the slope is nearly zero. We will model this covariate using the same type of time varying model proposed for chf. Before we **stsplit** the data into time intervals of constant $d_1(t)$ we fit the four covariate PH model.

CHAPTER 9 SOLUTIONS

```
. stcox age sex chf miord01 , nolog nohr noshow

Cox regression -- Breslow method for ties

No. of subjects =         481              Number of obs   =       481
No. of failures =         249
Time at risk    =    2284.887068
                                           LR chi2(4)      =    112.96
Log likelihood  =    -1364.1276            Prob > chi2     =    0.0000

------------------------------------------------------------------------------
         _t |
         _d |      Coef.   Std. Err.      z    P>|z|     [95% Conf. Interval]
------------+-----------------------------------------------------------------
        age |   .033594    .0058681     5.72   0.000     .0220927    .0450953
        sex |   .064168    .1328668     0.48   0.629    -.1962461    .3245822
        chf |  .7674217    .1363346     5.63   0.000     .5002107    1.034633
    miord01 |  .3651857    .1286573     2.84   0.005     .1130219    .6173494
------------------------------------------------------------------------------
```

```
. stsplit d1 , at(600)
(316 observations (episodes) created)

. tab d1

         d1 |      Freq.     Percent        Cum.
------------+-----------------------------------
          0 |        481       60.35       60.35
        600 |        316       39.65      100.00
------------+-----------------------------------
      Total |        797      100.00

. replace d1 = 1 if d1==600
(316 real changes made)
```

```
. stcox age sex chf miord01 , nolog nohr noshow   * check to see we get the same
base model*

Cox regression -- Breslow method for ties

No. of subjects  =         481                  Number of obs    =         797
No. of failures  =         249
Time at risk     =      834555
                                                 LR chi2(4)       =      112.96
Log likelihood   =  -1364.1276                   Prob > chi2      =      0.0000

------------------------------------------------------------------------------
     _t |
     _d |     Coef.    Std. Err.      z     P>|z|    [95% Conf. Interval]
--------+---------------------------------------------------------------------
    age |   .033594    .0058681     5.72   0.000    .0220927    .0450953
    sex |   .064168    .1328668     0.48   0.629   -.1962461    .3245822
    chf |  .7674217    .1363346     5.63   0.000    .5002107    1.034633
 miord01|  .3651857    .1286573     2.84   0.005    .1130219    .6173494
------------------------------------------------------------------------------
```

. gen chfearly=chf*(1-d1)

. gen chflate=chf*d1

. gen mioearly = miord01*(1-d1)

. gen miolate = miord01*d1

```
. stcox age sex chfearly chflate mioearly miolate , nolog nohr noshow

Cox regression -- Breslow method for ties

No. of subjects  =         481                  Number of obs    =         797
No. of failures  =         249
Time at risk     =      834555
                                                 LR chi2(6)       =      124.68
Log likelihood   =  -1358.2659                   Prob > chi2      =      0.0000

------------------------------------------------------------------------------
      _t |
      _d |     Coef.    Std. Err.      z     P>|z|    [95% Conf. Interval]
---------+--------------------------------------------------------------------
     age |  .0335521    .0058693    5.72   0.000    .0220485    .0450558
     sex |  .0459072    .1333127    0.34   0.731   -.2153809    .3071953
 chfearly|  .9969967    .1730007    5.76   0.000    .6579215    1.336072
  chflate|  .3263029    .2272277    1.44   0.151   -.1190551    .771661
 mioearly|   .561091    .1571987    3.57   0.000    .2529872    .8691949
  miolate| -.0877121     .23815    -0.37   0.713   -.5544774    .3790532
------------------------------------------------------------------------------
```

The *p*-values for the Wald statistics for the time varying covariates support early but no late effects for both chf and miord.

Thus as was the case for the UIS study
1. The plots from fitting the Aalen model have helped identify time varying effects
2. The PH model with the time varying effects provides sharper and more realistic estimates of effect.

In practice, one question that would need to be resolved is the clinical plausibility of the choice of cut points for the time varying effects in both the UIS and WHAS studies.